SpringerBriefs in Earth System Sciences

Series Editors

Gerrit Lohmann, Universität Bremen, Bremen, Germany

Lawrence A. Mysak, Department of Atmospheric and Oceanic Science, McGill University, Montreal, QC, Canada

Justus Notholt, Institute of Environmental Physics, University of Bremen, Bremen, Germany

Jorge Rabassa, Labaratorio de Geomorfología y Cuaternar, CADIC-CONICET, Ushuaia, Tierra del Fuego, Argentina

Vikram Unnithan, Department of Earth and Space Sciences, Jacobs University Bremen, Bremen, Germany

SpringerBriefs in Earth System Sciences present concise summaries of cutting-edge research and practical applications. The series focuses on interdisciplinary research linking the lithosphere, atmosphere, biosphere, cryosphere, and hydrosphere building the system earth. It publishes peer-reviewed monographs under the editorial supervision of an international advisory board with the aim to publish 8 to 12 weeks after acceptance. Featuring compact volumes of 50 to 125 pages (approx. 20,000—70,000 words), the series covers a range of content from professional to academic such as:

- A timely reports of state-of-the art analytical techniques
- bridges between new research results
- snapshots of hot and/or emerging topics
- literature reviews
- in-depth case studies

Briefs are published as part of Springer's eBook collection, with millions of users worldwide. In addition, Briefs are available for individual print and electronic purchase. Briefs are characterized by fast, global electronic dissemination, standard publishing contracts, easy-to-use manuscript preparation and formatting guidelines, and expedited production schedules.

Both solicited and unsolicited manuscripts are considered for publication in this series.

Gaël Kermarrec · Vibeke Skytt · Tor Dokken

Optimal Surface Fitting of Point Clouds Using Local Refinement

Application to GIS Data

 Springer

Gaël Kermarrec
Institute for Meteorology and Climatology
Leibniz University Hannover
Hanover, Germany

Vibeke Skytt
Mathematics and Cybernetics
SINTEF Digital
Oslo, Norway

Tor Dokken
Mathematics and Cybernetics
SINTEF Digital
Oslo, Norway

ISSN 2191-589X   ISSN 2191-5903 (electronic)
SpringerBriefs in Earth System Sciences
ISBN 978-3-031-16953-3   ISBN 978-3-031-16954-0 (eBook)
https://doi.org/10.1007/978-3-031-16954-0

This Springer imprint is published by the registered company Springer Nature Switzerland AG
The registered company address is: Gewerbestrasse 11, 6330 Cham, Switzerland

Foreword

Splines are piecewise polynomial functions and are ubiquitous in the sciences. They play a prominent role in computer-aided geometric design, signal processing, data analysis, visualization, numerical simulation, probability, and many more.

In one variable, splines are usually represented in terms of so-called B-splines. B-splines turn out to be the most useful spline basis functions because they possess several properties that are important from both theoretical and computational point of view. Even though the concept of B-splines was already known before, the modern B-spline theory roots in the seminal works by Isaac Jacob Schoenberg in the mid-twentieth century and has many important developments ever since. When moving to multiple variables, tensor-product B-spline representations are the most common choice thanks to their simplicity of construction and the inheritance of all nice properties of the univariate representations. But they have limitations, and therefore, several alternative spline technologies have been proposed more recently for dealing with more general domains, unstructured partitions, and local refinement.

Oslo has a long-standing tradition in spline research, with several fundamental contributions. The so-called Oslo algorithm (1980) is probably one of the most famous outcomes of this research, a general method for inserting degrees of freedom to a given B-spline curve. A recent contribution is the creation of "LR B-splines" (2013), an elegant theory for performing local refinement on B-spline surfaces or volumes. Tor Dokken and the Geometry Group at SINTEF are a driving force behind the development and application of LR B-splines and their promotion within both academia and industry. His endless enthusiasm and encouragement are a great stimulus for all researchers who are working—or will work—on the topic.

This book will introduce you in the world of LR B-splines, with a particular focus on their application to surface fitting. Finding a compact surface description of scattered observations is an important problem. LR B-splines could provide an excellent tool for this. Adaptive refinement allows for local updates of the LR B-spline surface where the distance to the point cloud exceeds a prescribed tolerance, without introducing a huge number of redundant degrees of freedom. The theoretical framework is illustrated with practical examples of bathymetry and landslides GIS datasets. It is a comprehensive source for everyone interested in adaptive surface fitting and splines.

Rome, Italy   Hendrik Speleers
April 2022

Acknowledgements

Dr. Gaël Kermarrec is supported by the Deutsche Forschungsgemeinschaft under the project KE2453/2-1. This contribution proposes innovative solutions to perform the surface approximation of noisy and correlated point clouds from terrestrial laser scanner, with the goal to improve spatio-temporal monitoring of object changes based on mathematical surfaces. This publication was financially supported by the Leibniz University Hannover's Open Access Publishing Fund.

The work of Chief Scientist Dr. Tor Dokken (SINTEF) and Senior Scientist Vibeke Skytt (SINTEF) on LR B-splines for the representation, processing, and analysis of big GIS point clouds presented in this SpringerBrief has been financially supported by the EU through the FP7 project IQmulus (FP7-ICT-2011-318787) (2012–2016) and by the Research Council of Norway through the IKTPluss project ANALYST (Contract no. 270922)(2017–2021).

The authors would like to express their most sincere thanks to Daniel Czerwonka-Schröder for having shared the landslide dataset used in Chap. 6. This measurement campaign was supported by the Research Fund of the European Union for Coal and Steel [RFCS project number 800689 (2018)]. Ke Lu is further gratefully thanked for having implemented and performed some simulations on the AIC in Chap. 4.

We also want to warmly thank Henning Sundby from the Norwegian Hydrographic Service (a division of the Norwegian Mapping Authority) for pointing us which datasets to use for testing the spline technology presented in this SpringerBrief. He has also been an excellent dialogue partner with respect to the challenges and possible technological solutions for handing of large ocean floor and sea bottom datasets, as well as the use of spline technologies to address these challenges.

Senior Scientist Dr. Oliver Barrowclought from SINTEF has been a driving force for improving the text of this SpringerBrief, for which he deserves a warm thanks.

Highlights of the SpringerBrief

1. Locally Refined B-splines surfaces approximate huge, scattered, and noisy point clouds efficiently
2. The surface approximation is adaptive and allows for local refinement of various point clouds as, e.g. LIDAR or sonar datasets
3. LR B-spline surfaces and volumes are ideal for spatio-temporal analysis with geomorphological or geodetic applications
4. The format of LR B-spline surfaces can be made compatible with GIS software to promote interoperability. The C++ functions are freely available

About This Book

With the development of high-rate sensors based on light detection and ranging technologies, geospatial data representing terrains and seabeds contains millions of points. Performing a surface approximation is an efficient way to reduce, smooth, and structure the recorded data. Prominent applications are the analysis of spatio-temporal deformation, or the drawing of contour lines.

In this SpringerBrief, we dive into the concept of Locally Refined (LR) B-splines to approximate point clouds. We describe both intuitively and mathematically how local adaptive refinement is performed and highlight its advantages over other methods. Various examples using datasets from a sonar, a terrestrial laser scanner, and a hand-scanner illustrate the methodology. A suitable procedure to deal with outliers and voids within the context of surface approximation is proposed. We conclude by highlighting the potential of LR B-splines surfaces and volumes to perform spatio-temporal geomorphological or geodetic deformation analysis, as promising applications.

This SpringerBrief is written for a wide audience: from a practitioner wishing to perform surface approximation of point clouds, to mathematicians interested in understanding the concepts of local refinement and their potential applications.

Contents

About the Authors

Dr. Gaël Kermarrec received her M.S. (2000) in geodesy from the Ecole Nationale des Sciences Geographiques in Marne la Vallee, France. She made her promotion at the Karlsruhe Institute of Technology (KIT) in Germany in 2016, with a focus on GPS correlations modelling. Her research topics are surface modelling of point clouds from terrestrial laser scanners, as well as correlation analysis. Her main field of interest is physical modelling of correlations due to atmospheric turbulence.

Vibeke Skytt M.Sc. graduated from the University of Oslo in 1986 with a Master of Science in numerical analysis and has since then worked as a researcher and senior researcher at SINTEF in Oslo. Her field is geometry with emphasis on splines, with applications in various areas such as CAD, reverse engineering, and modelling for isogeometric analysis. She has, in recent years, worked with representation of geographic data with locally refined splines.

Dr. Tor Dokken received his Dr. Philos degree from the University of Oslo in 1997. He is Chief Scientist and Research Manager for the Geometry Group in SINTEF Digital, has been actively participating in the SIAM Activity Group for Geometric Design, and is one of the organizers of the Dagstuhl Seminars on Geometric Modelling: Interoperability and New Challenge. Coordinating a series of EU-funded projects addressing design and simulation, isogeometric analysis, additive manufacturing, and GIS Big Data is and has been one of his key activities. He is, in the period 2020–2024, the coordinator for the H2020 Innovation Action Change2Twin addressing digital twins for manufacturing and part of a parallel Innovation Action PULSATE addressing advanced laser based and additive manufacturing.

Acronyms

AIC	Akaike Information Criterion
ASCII	American Standard Code For Information Interchange
BIC	Bayesian Information Criterion
CAD	Computer-Aided Design
CT	Computational time
DEM	Digital Elevation Model
GIS	Geographic Information System
HB	Hierarchical B-splines
IDW	Inverse Distance Weighting
IgA	Isogeometric Analysis
IQR	Interquartile Range
LIDAR	Light Detection And Ranging
LR	Locally Refined
LS	Least Squares
MAE	Mean Average Distance
MBA	Multilevel B-spline Approximation
NURBS	Non-Uniform Rational B-splines
RBF	Radial Basis Functions
RMSE	Root Mean Squared Error
STL	Standard Triangle Language
THB	Truncated Hierarchical B-splines
TIN	Triangulated Irregular Network
TLS	Terrestrial Laser Scanner
TP	Tensor Product
UMB	Univariate Minimal Support B-spline Basis

Chapter 1
Introduction

Abstract With the development of high rate sensors based on LIDAR (light detection and ranging) and sonar technology, geospatial data representing terrain or seabed often contains millions of points. Performing a surface approximation of the point clouds is an elegant way to reduce noisy and unorganized data to a mathematical surface with just a few coefficients to estimate. Traditional spline surfaces are able to compactly represent smooth shapes, but lack the ability to adapt the representation locally to the point clouds. Locally Refined (LR) B-spline surfaces address that challenge as they have the nice property of being locally refinable. Their format can be made compatible with most Geographic Information System (GIS) software, and they facilitate various applications such as the drawing of contour lines or spatio-temporal deformation analysis. This introduction aims to explain the need for surface approximation, and present the state of the art in that domain. We compare the LR B-spline approach with different methods for surface approximation including raster, and triangular irregular networks.

Keywords Geospatial data · LR B-Spline surfaces · Approximation · Surfaces in GIS

1.1 The Why and How of Surface Approximation

The advance of contactless laser range scanners, either terrestrial, airborne or underwater as well as sonars, enables to capture 3D data of large areas rapidly and with a high accuracy [Weh99, Eno19]. The applications of such sensors are diverse, going from forest inventory to agricultural monitoring, deformation analysis of bridges and dams, underwater or seafloor shell fragment characterization but also cultural heritage, to name only but a few (see, e.g., [Muk16] or [Wu22] for a review of applications). While working directly with the recorded point clouds may be adequate for visualization, or animation purposes, the manipulation of millions of points is less attractive as soon as shape analysis is needed [Flo02]. Within the context of geospatial data approximation or reverse engineering [Raj08], it is favorable to convert the observations to a mathematical surface. Here the latter are defined by parametric

© The Author(s) 2023

G. Kermarrec et al., *Optimal Surface Fitting of Point Clouds Using Local Refinement*,
SpringerBriefs in Earth System Sciences,
https://doi.org/10.1007/978-3-031-16954-0_1

equations and approximate data by minimizing the distance between the point clouds and the approximated surface.

Throughout the book, we define an "approximation" as a counterpart to mathematical "interpolation" for which the resulting surface passes through all the data points (see, e.g., [Fol86]). In the context of Geographic Information Systems (GIS), the term "spatial interpolation" is given a more broad definition. It is the process of using points with known values to estimate values at other points. Spatial interpolation can further be divided into exact or inexact interpolation. We will reserve the term "interpolation" to mean an exact fitting of a set of data points while inexact fitting will be denoted "approximation". Interpolation is unfavorable when a huge number of points is available. The approximation of noisy, unstructured and scattered point clouds **transforms data to information**: The resulting surfaces are less redundant and complex than when interpolation is performed.

Rigorous statistical testing of the deformation of objects such as bridges, dams or tunnels with underlying safety applications, can be best performed with mathematical surfaces. Within a geodetic context, they further make huge point clouds easy to handle and manipulate. Unfortunately, many practitioners are hesitating to use parametric surfaces to approximate their data, expressing concerns such as *"is it accurate enough?"*, *"is it time consuming?"*, *"I don't understand formulas"*. Thus, the use of mathematical approximations of point clouds can only grow if easy-to-use and easy-to-understand approximation methods are proposed. The following criteria have to be considered:

1. **Accuracy**: The error between the fitted and original data set should be kept small.
2. **Smoothness**: Surface ripples due to the approximation of outliers or voids should be avoided.
3. **Conciseness**: The resulting surface should contain as few parameters as possible.
4. Automaticity and **interoperability** with existing GIS software: the format should be flexible.
5. **Computational time**: The processing has to be fast. It is one of the major requests from a practitioner perspective, and slowness strongly deters the use of parametric modelling.

The approach we follow in this SpringerBrief is to use Locally Refined B-spline surfaces, abbreviated as LR B-spline surfaces, for approximating geospatial data [Dok13, Sky15].

1.2 Surface Representation of Geospatial Data

There are many other data representations in addition to LR B-spline surfaces that can be used to approximate point clouds. In the following sections, we present two prominent examples: Raster representation and triangulated irregular network (TIN). We provide a short comparison of these methods with LR B-spline surface fitting.

Other representations such as radial basis functions used, e.g., for gravity field modelization [Ten08] and trend surfaces are not directly applicable for approximation of
large datasets and therefore omitted. We note that in GIS the term "spline" most often
refers to splines in tension or regularized splines, which differ from tensor product
(TP) B-splines surfaces. In Computer Aided Design (CAD) a rational version of TP
B-spline surfaces is used, the so-called Non Uniform Rational B-splines (NURBS)
surfaces.

1.2.1   Raster Representation

The raster representation is the most frequently used data format in GIS, [Bis18].
The digital elevation model (DEM) is often represented as a raster. The raster is an
approximate representation as the scattered input data are not exactly fitted. A given
cell contains a single value, often the elevation, so that the level of detail is restricted
by the raster cell resolution. If this resolution is low compared to the variation in the
data, meaning that there is a large height difference between points in a cell, then
the result may be inaccurate. On the contrary, if the resolution is too high, the data
volume grows more than necessary. A trade-off between accuracy and data volume
must be made, and this is particularly mandatory when there are large differences
in the local variation of the data in different areas. However, the raster remains a
compact, highly structured and efficient representation. Proposals have been made
based on a compact data structure to access a given datum or portion of the data more
rapidly, [Sil21].

Figure 1.1a shows a cloud of 999,751 points consisting of classified ground points
and points from the sea surface. A raster representation with 1 m resolution is shown in
Fig. 1.1b. The raster is computed using inverse distance weighting (IDW, [She68]).
It has size 800×600 and the average number of points used for estimating the

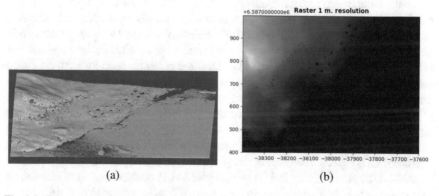

(a)   (b)

Fig. 1.1   Raster representation of terrain. **a** Point cloud from Fjøløy in Norway, **b** raster representation visualized with Python Rasterio

Fig. 1.2 A small
triangulated point set. The
data points are red and the
triangulation edges are
shown in blue

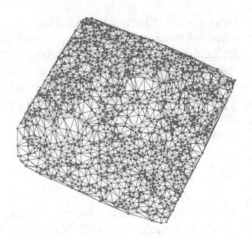

raster points is 5676.79. The accuracy of this raster representation is addressed in
Sect. 1.2.5.1.

Several approaches are available to estimate values between the existing samples.
We mention the selection in the cell center or the bivariate evaluation. Alternatively, the estimated value can be computed from a bivariate surface interpolating
the four surrounding sample values. Kriging or IDW can be also used in this context,
see [Oli90] or [She68]. The reader is referred to, e.g., [Wis11], [Mit05] or [Fis06]
for more details and specific comparisons between methods.

1.2.2 Triangulated Irregular Network (TIN)

A TIN is a continuous surface representation frequently used in GIS. This is a flexible
format for geospatial data that allows adaptation to local variations, and is highly
accurate. Similarly to raster representations and LR B-splines, an approximation is
required to restrict the data size. The nodes of a TIN are distributed variably to create
an accurate representation of the terrain. TINs can, thus, have a higher resolution in
areas where a surface is highly variable or where more detail is desired and a lower
resolution in areas that are less variable. They are typically used for high-precision
modelling of smaller areas. In [Nel94] a triangulated surface is used to represent
drainage-basins while hydrological similarity is used in the TIN creation in [Viv04].

TINs have a more complex data structure than raster surfaces and tend to be more
expensive to build and process. Points in-between the corner points in a triangle are
calculated by linear interpolation. This can give a jagged appearance of the surface.
The problem is especially visible at sharp or nearly sharp edges, but can be remedied
by methods like constrained Delaunay triangulation. A comprehensive discussion on
various aspects with triangulation in terrain modelling is presented in [Li09].

The triangulation shown in Fig. 1.2 is interpolating a sparse set of terrestrial data points and created with unconstrained Delaunay triangulation. The data is fetched from LIDAR measurements of the island Fjøløy in Norway and is subsampled to improve visibility.

1.2.3 B-Spline Curves and Tensor Product Surfaces

B-spline curves are piecewise polynomial curves with continuity between adjacent polynomial pieces embedded in the curve formulation. The joints between the polynomial pieces are defined by the so-called knot vector. The polynomial degree can be chosen but is often selected to be three. A B-spline curve is described as a linear combination of a set of coefficients and corresponding B-splines basis functions, see [Pie91]. The maximum possible continuity between the polynomial pieces is equal to the polynomial degree minus one.

The B-spline basis functions are themselves piecewise polynomials and have several attractive properties:

- Non-negativity
- Partition of unity (the B-splines in a given parameter always sum up to one)
- Linear independence
- Limited support, i.e., the values of the basis functions are different from zero in a limited interval given by the knot vector, implying that a modification of one surface coefficient will change the curve only locally.

The properties of the B-splines imply that the representation is numerically stable and that a B-spline curve is bounded by its coefficients. The curve will always lie inside the polygon described by its coefficients. A B-spline curve is locally refinable, i.e., new knots can be inserted into the curve description as required.

A bivariate tensor product (TP) B-spline surface is constructed by taking the tensor product of the basis functions defined over knot vectors in two parameter directions, and defining appropriate coefficients. This construction carries over the attractive properties of non-negativity, limited support of the B-spline basis functions, partition of unity and linear independence. Unfortunately, the TP B-spline surface formulation does not allow local refinement. If a new knot is entered in one of the parameter directions, it will cover the entire parameter domain in the other direction.

1.2.4 Locally Refined B-Spline Surfaces

The LR B-spline surface [Dok13] is one approach to solve the problem of lack of local refinability of TP B-spline surfaces. Other approaches include hierarchical B-splines [For88, Bra18], its variation called Truncated Hierarchical B-splines [Gia12], and T-splines [Sed03]. For LR B-splines, the starting point is always a TP B-spline surface with a corresponding mesh of lines defined by the knots in the

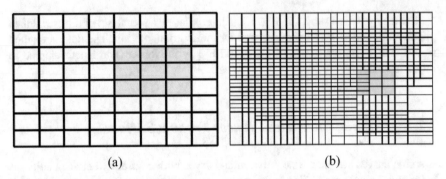

Fig. 1.3 TP and LR meshes. **a** The initial TP mesh with the support of one biquadratic B-spline highlighted, **b** LR mesh after insertion of several meshlines with the support of one biquadratic B-spline highlighted

two parameter directions (a TP mesh). The TP mesh is converted into an LR mesh and new meshlines are inserted into the mesh to refine the surface. The surface coefficients are updated accordingly. The new meshlines do not need to cover the entire region, but must traverse the support of at least one B-spline. Each new meshline leads to one or more B-splines being split to give rise to more B-splines and consequently more approximation freedom. The TP mesh is a special case of an LR mesh. Figure 1.3 illustrates how a TP mesh can be extended to an LR mesh through multiple mesh refinements.

For LR B-spline surfaces the construction of the TP B-splines spanning the spline space is similar to the construction of the TP B-splines spanning the spline space of TP B-spline surfaces. In both cases they are TP B-splines that are defined from a subset of the knot vectors in the two parameter directions. The TP B-splines are regular except for possible variation in the width and height of mesh cells due to varying intervals between adjacent knots. The LR B-splines can differ greatly both in terms of the size of the support and the number of LR B-splines overlapping a particular parameter point in the surface domain. The LR B-splines are non-negative, have limited support, and possess the partition of unity property. The collection of LR B-splines are not linearly independent by default, but possible occurrences of linear dependency can be detected and removed.

Figure 1.4a shows a biquadratic LR B-spline surface approximating the point cloud in Fig. 1.1 (tolerance 0.5 m and 5 iteration steps in an adaptive surface fitting algorithm, see Chap. 3). Here the advantages of local refinement are highlighted:

1. The surface is locally refined by the algorithm where the accuracy requirement is not met.
2. The final LR mesh in Fig. 1.4b is considerably more dense in the steep areas of the data set than where the surface represents the sea surface.

The accuracy of this surface representation will be addressed in Sect. 1.2.5.1.

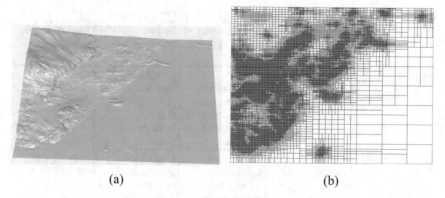

(a)   (b)

Fig. 1.4 LR B-spline representation of the point cloud shown in Fig. 1.1. **a** The approximating surface, **b** the associated LR mesh

1.2.5 Comparison Between Approximation Strategies

Table 1.1 provides an overview over the surface representations described previously. TINs and LR B-spline surfaces are created by adaptive algorithms, so that the degrees of freedom in the surface can be determined according to the need, and the accuracy of the fit is directly available. The raster representation and the TP B-spline surfaces—a generalization of the raster representation—are less flexible. The TP B-spline surface description is more flexible than the raster due to the choice of polynomial bidegree and/or variable knot vectors. However, the raster representation is slightly more compact than the TP B-spline surface for the representation of a piecewise constant or piecewise bilinear function. The TP B-spline and LR B-spline methods provide smoother surfaces than the other methods. The lack of local refinement, however, implies that the TP B-spline surface size grows much faster compared with LR B-spline surfaces.

1.2.5.1 Comparison Raster/LR B-Spline Surfaces

Here we wish to point out the advantages of an LR B-spline surface approximation with respect to the raster approximation. To that end, we come back to the point cloud approximated in Sects. 1.2.1 and 1.2.4. Figure 1.5 shows the point cloud coloured according to the distance to (a) the approximated raster surface, and (b) the LR B-spline surface. The raster surfaces were created with IDW and evaluated by linear interpolation. This is not necessarily the optimal approximation method, but it is a method frequently used in GIS. Visually, the distance between the point cloud and the surface is largest for the raster surface: The most distant points are concentrated in areas with much shape variation in the terrain whereas the sea surface is accurately represented in both cases. These results are summarized in Table 1.2. The file sizes

Table 1.1 Summary of surface formats for representing terrain and seabed

Surface type	Representation and data structure	Algorithm and control of accuracy	Surface smoothness	Restricting data volume
Raster	Values on regular mesh	Spatial interpolation to define sample values. Accuracy checked after creation	Depends on interpolation method for evaluation between mesh points	Pre-set mesh resolution defining data volume
B-spline surface	Piecewise polynomials on regular mesh, any bidegree	Coefficients calculated by local/global approximation. Accuracy checked after surface creation	Depends on polynomial bidegree and knot multiplicity	Pre-set mesh resolution defining data volume
TIN	Triangulation	Triangulate point cloud + thinning or adaptive triangulation. Accuracy can be checked during creation	Piecewise linear	Pre-set approximation tolerance and/or max allowed data volume
LR B-spline surface	Piecewise polynomials on LR axis-parallel mesh, any bidegree	Coefficients calculated by local/global approximation. Local adaption by checking accuracy during construction and refinement where needed.	Depends on polynomial bidegree and knot multiplicity	Pre-set approximation tolerance or restrictions in adaptive algorithm

of the raster surfaces (GeoTIFF) are generally larger than for LR B-spline surface (ASCII), so are the distances between the points and the surface. This data set favours the LR B-spline surface with one part representing the horizontal sea surface and one part a terrain with considerable height variations. The property to allow for local variations is the main advantage of LR B-spline surface approximation, not to mention the strong reduced data size of the final surface.

To summarize, LR B-spline surfaces:

1. Allow a smooth representation of point clouds,
2. Avoid the ragged appearance that can occur for TIN,
3. Are particularly advantageous in terms of the number of coefficients to estimate for fitting huge terrain and seabed data.

(a)   (b)

Fig. 1.5 The point cloud from Fig. 1.1, **a** coloured according to the distance to the raster surface with 1 m resolution shown in Fig. 1.1, **b** the LR B-spline surface in Fig. 1.4. The size of the most distant points is increased compared to points closer to the surface for improved visibility. White points lie closer to the surface than 0.5 m, green points lie below the surface and red points above

Table 1.2 Comparison between raster representation and LR B-spline surface

Surface	File (MB)	Max $\|d\|$	MAE	$\|d\| > 3$	$0.5 < \|d\| < 3$ (%)	$\|d\| < 0.5$ (%)
Raster, 1 m	1.9	13.498	0.116	0.02%	4.1	95.9
Raster, 0.5 m	7.4	11.99	0.073	0.004%	1.3	98.7
LR, 5 steps	1.2	3.23	0.065	1 point	0.6	99.4
LR, 12 steps	3.1	0.5	0.055	0%	0	100

Raster surfaces are computed with 0.5 and 1 m resolution, LR B-spline surfaces with 5 and 12 iteration steps using a tolerance of 0.5 m. Distances are given in m and point distribution in percentage of the total number of points (1,643,865). MAE = average absolute value of distance, and $\|d\|$ is the absolute value of the distance between a point and the surface

We will highlight these properties in Chaps. 5 and 6.

1.2.5.2 Summary

Many applications can be derived from approximation of point clouds with mathematical surfaces, such as the drawing of contour lines, or rigorous deformation analysis based on statistical tests. The result of the approximation and the choice of the method depend on the characteristics of the data, the purpose of the surface generation and user defined criteria, such as the computational time. For GIS applications, the LR B-spline surface with adaptive local refinement is favourable and the principle intuitive and understandable ([Sky22] for some examples, [Ker21] for geodetic applications). We point out that neither LR B-spline surfaces, nor raster nor TIN is the most appropriate representation: This latter does not exist. The definition of goodness of fit depends on the applications and the data at hand.

1.3 Reminder of the Present SpringerBrief

In Chap. 2, we will present in details the concepts of LR B-splines and review alternative local approximation methods, such as hierarchical B-splines and T-splines. In the LR B-spline surface approximation with adaptive refinement, parameters are inserted locally, when needed. For geodetic objects such as a bridge or for landslides, this approach is favourable as more details can be needed in domains where, e.g., strong deformations are likely to happen or the object has edges.

The procedure of adaptive refinement will be developed in Chap. 3. The algorithms are optimized for a wide use within a GIS or geodetic context. For approximating geospatial points, biquadratic surfaces have proven to be a good choice [Sky22]. They provide a good balance between smoothness and flexibility. Once an LR B-spline surface representation of some scattered data is obtained, terrain information can be derived such as, e.g., contour curves, slope and aspect ratio. Deformation analysis and statistical tests can be performed at the level of the surface approximation [Sky22, Ker20, Ker21]. We will develop such applications in Chaps. 5 and 6 and present how specific challenges such as data gaps and outliers can be handled efficiently. The concept of LR B-splines volumes will be described. The optimal determination of approximation parameters such as the tolerance with respect to the noise level of the point cloud, is part of Chap. 4, which addresses how to choose less empirically some refinement parameters or strategies.

References

[Bis18] Bishop, M. P., Brennan, W. Y., & Huo, D. (2018). Geomorphometry: Quantitative landsurface analysis and modeling. In *Reference module in earth systems and environmental sciences*. ISBN 9780124095489, https://doi.org/10.1016/B978-0-12-409548-9.11469-1

[Bra18] Bracco, C., Giannelli, C., Großmann, D., & Sestini, A. (2018). Adaptive fitting with THB-splines: Error analysis and industrial applications. *Computer Aided Geometric Design*. https://doi.org/10.1016/j.cagd.2018.03.026

[Dok13] Dokken, T., Pettersen, K. F., & Lyche, T. (2013). Polynomial splines over locally refined boxpartitions. *Computer Aided Geometric Design*. https://doi.org/10.1016/j.cagd.2012.12.005

[Eno19] Enomoto, E., et al. (2019). Application of range finder by image sensor in the underwater environment. *2019 International Conference on Electron*. https://doi.org/10.23919/ELINFOCOM.2019.8706398

[Fis06] Fisher, P. F., & Tate N. J. (2006). Causes and consequences of error in digital elevation models. *Progress in Physical Geography*. https://doi.org/10.1191/0309133306pp492ra

[Flo02] De Floriani, L., & Spagnuolo, M. (2002). *Shape interrogation for computer aided design and manufacturing*. Springer.

[Flo05] Floater, M. S., & Hormann, K. (2005). Surface parameterization: A tutorial and survey. In N. A. Dodgson, M. S. Floater, & M. A. Sabin (Eds.), *Advances in multiresolution for geometric modelling, mathematics and visualization*. Springer.

[Fol86] Foley, T. A. (1986). Scattered data interpolation and approximation with error bounds. *Computer Aided Geometric Design*. https://doi.org/10.1016/0167-8396(86)90034-8

[For88] Forsey, D. R., & Bartels, R. H. (1988). Hierarchical B-spline refinement. In *SIGGRAPH 88 Conference Proceedings* (Vol. 4, pp. 205–212).

[Fra82] Franke, R. (1982). Scattered data interpolation: Tests of some methods. *Mathematics of Computation.* https://doi.org/10.2307/2007474

[Gia12] Giannelli, C., Jüttler, B., & Speleers, H. (2012). THB-splines: The truncated basis for hierarchical splines. *Computer Aided Geometric Design.* https://doi.org/10.1016/j.cagd.2012.03.025

[Kel19] Keller, W., & Borkowski, A. (2019). Thin plate spline interpolation. *The Journal of Geodesy.* https://doi.org/10.1007/s00190-019-01240-2

[Ker20] Kermarrec, G., Kargoll, B., & Alkhatib, H. (2020). Deformation analysis using B-spline surface with correlated terrestrial laser scanner observations–A bridge under load. *Remote Sensing.* https://doi.org/10.3390/rs12050829

[Ker21] Kermarrec, G., Schild, N., & Hartmann, J. (2021). Fitting terrestrial laser scanner point clouds with t-splines: Local refinement strategy for rigid body motion. *Remote Sensing.* https://doi.org/10.3390/rs13132494

[Li09] Li, Y., & Yang, L. (2009). Based on delaunay triangulation DEM of terrain model. *Computer and Information Science.* https://doi.org/10.5539/cis.v2n2p137

[Mit05] Mitas, L., & Mitasova, H. (2005). Spatial interpolation. In P. Longley, M. F. Goodchild, D. J. Maguire, & D. W. Rhind (Eds.), *Geographic information systems—Principles, techniques, management, and applications* (pp. 481–498).

[Muk16] Mukupa, W., Roberts, G. W., Hancock, C. M., & Al-Manasir, K. (2016). A review of the use of terrestrial laser scanning application for change detection and deformation monitoring of structures. *Survey Review.* https://doi.org/10.1080/00396265.2015.1133039

[Nel94] Nelson, E. J., Jones, N. L., & Miller, A. W. (1994). Algorithm for precise drainage-basin delineation. *Journal of Hydraulic Engineering.* https://doi.org/10.1061/(ASCE)0733-9429(1994)120:3(298)

[Oli90] Oliver, M. A., & Webster, R. (1990). Kriging: a method of interpolation for geographical information system. *International Journal of Geographical Information Systems, 4*(3), 323–332.

[Pie91] Piegl, L. (1991). On NURBS: A survey. *IEEE Computer Graphics and Applications.* https://doi.org/10.1109/38.67702

[Raj08] Raja, V., & Fernandes, K. J. (2008). *Reverse engineering: An industrial perspective,* Springer series in advanced manufacturing. Springer.

[Sed03] Sederberg, T. W., Zheng, J., Bakenov, A., & Nasri, A. (2003). T-splines and T-NURCCs. *ACM Transactions on Graphics.* https://doi.org/10.1145/882262.882295

[She68] Shepard, D. (1968). A two-dimensional interpolation function for irregularly spaced data. In *Proceedings of 23rd National Conference, ACM,* pp. 517–523.

[Sky15] Skytt, V., Barrowclough, O., & Dokken, T. (2015). Locally refined spline surfaces for representation of terrain data. *Computers & Graphics.* https://doi.org/10.1016/j.cag.2015.03.006

[Sky22] Skytt, V., & Dokken, T. (2022). Scattered data approximation by LR B-spline surfaces. A study on refinement strategies for efficient approximation. In C. Manni, & H. Speleers (Eds.), *Geometric challenges in isogeometric analysis* (Vol. 49). Springer INdAM Series.

[Sil21] Silva-Coira, F., Paramá, J. R., de Bernardo, G., & Seco, D. (2021). Space-efficient representations of raster time series. *Information Sciences.* https://doi.org/10.1016/j.ins.2021.03.035

[Ten08] Tenzer, R., & Klees, R. (2008). *The choice of the spherical radial basis functions in local gravity field modeling.* https://doi.org/10.1007/s11200-008-0022-2

[Viv04] Vivoni, E. R., Ivanov, V. Y., Bras, R. L., & Entakhabi, D. (2004). Triangulated irregular networks based on hydrological similarity. *Journal of Hydrologic Engineering.* https://doi.org/10.1061/(ASCE)1084-0699(2004)9:4(288)

[Weh99] Wehr, A., & Lohr, U. (1999). Airborne laser scanning–An introduction and overview. *ISPRS Journal of Photogrammetry and Remote Sensing.* https://doi.org/10.1016/S0924-2716(99)00011-8

[Wis11] Wise, S. (2011). Cross-validation as a means of investigating DEM interpolation error. *Computers & Geosciences.* https://doi.org/10.1016/j.cageo.2010.12.002

[Wu22] Wu, C., Yuan, Y., Tang, Y., & Tian, B. (2022). Application of terrestrial laser scanning (TLS) in the architecture, engineering and construction (AEC) industry. *Sensors*. https://doi.org/10.3390/s22010265

Chapter 2
Locally Refined B-Splines

Abstract The univariate minimal support B-spline basis (UMB) has been used in Computer Aided Design (CAD) since the 1970s. Freeform curves use UMB, while sculptured surfaces are represented using a tensor product of two UMBs. The coefficients of a B-spline curve and surface are respectively represented in a vector and a rectangular grid. In CAD-intersection algorithms for UMB represented objects, a divide-and-conquer strategy is often used. Refinement by knot insertion is used to split the objects intersected into objects of the same type with a smaller geometric extent. In many cases the intersection of the resulting sub-objects has simpler topology than the original problem. The sub-objects created are represented using their parents' UMB format and deleted when the sub-problem is solved. Consequently, no global representations of the locally refined bases are needed. This is contrary to when locally refined splines are used for approximation of large point sets. As soon as a B-spline is locally refined, the regular structure of UMB objects in CAD is no longer valid. In this chapter we discuss how Locally Refined B-splines (LR B-splines) address this challenge and present the properties of LR B-splines.

Keywords Locally Refined B-splines · Minimal support basis · Refinement

2.1 Introduction

An introduction to B-splines, tensor product (TP) B-splines and the univariate minimal support B-spline basis (UMB) is provided in Sect. 2.2. LR B-splines are a generalization of local refinement of curves represented in UMB to the multivariate case, see Sect. 2.3 for details on refinement of B-spline curves. An overview of B-spline based methods for local refinement is addressed in Sect. 2.4, with LR B-splines described in more details in Sect. 2.5. To give a first motivation for the approach of LR B-splines, we will now illustrate the LR B-spline representation for the bivariate case, and introduce the notation used for describing LR B-splines.

© The Author(s) 2023

G. Kermarrec et al., *Optimal Surface Fitting of Point Clouds Using Local Refinement*,
SpringerBriefs in Earth System Sciences,
https://doi.org/10.1007/978-3-031-16954-0_2

We start from the traditional representation of a CAD-type TP B-spline surface.

$$F(u, v) = \sum_{i=0}^{N_1-1} \sum_{j=0}^{N_2-1} c_{i,j} B_{i,p_1}(u) B_{j,p_2}(v), \quad (u, v) \in [u_{p_1}, u_{N_1}] \times [v_{p_2}, v_{N_2}].$$

Here $c_{i,j} \in R^d, i = 0, \ldots, N_1 - 1, j = 0, \ldots, N_2 - 1$ are the surface coefficients and d is the dimension of the geometry space. Note that for parametric curves and surfaces, when the geometry space has dimension larger than 1 ($d > 1$), i.e., when we have a parametric curve or surface, the coefficients are frequently denoted control points. In this chapter we will use coefficients except for cases where the use of control points is necessary to ensure clarity. We also have two UMBs, one in each parameter direction: $B_{i,p_1}(u), i = 0, \ldots, N_1$ is an UMB defined over the knot vector $\{u_0, \ldots, u_{N_1+d_1}\}$, and $B_{j,p_2}(v), j = 0, \ldots, N_2$ is an UMB defined over the knot vector $\{v_0, \ldots, v_{N_2+d_2}\}$. Note that the TP B-splines $B_{i,j,p_1,p_2}(u, v) = B_{i,p_1}(u) B_{j,p_2}(v)$ are implicitly defined in the equation above. Making the TP B-splines explicit, the equation above can be reformulated to

$$F(u, v) = \sum_{i=0}^{N_1-1} \sum_{j=0}^{N_2-1} c_{i,j} B_{i,j,p_1,p_2}(u, v), \quad (u, v) \in [u_{p_1}, u_{N_1}] \times [v_{p_2}, v_{N_2}].$$

In the equation above, the grid structure of coefficients and TP B-splines is still present. To prepare for LR B-splines we have to reformulate the above to a collection of TP B-splines with corresponding coefficients. Now define

$$\mathcal{B}_0 = \{B \mid B = B_{i,j,p_1,p_2}(u, v), i = 0, \ldots, N_1 - 1, j = 0, \ldots, N_2 - 1\}.$$

This allows us to express $F(u, v)$ as

$$F(u, v) = \sum_{B \in \mathcal{B}_0} c_B s_B B(u, v).$$

The scaling factors s_B are introduced to allow scaling of refined B-splines to provide a scaled partition of unity of a collection of LR B-splines. For a TB-surface all $s_B \equiv 1$. Note that we index the coefficients c_B and the scaling factors s_B by the TP B-spline B that it corresponds to.

2.2 B-Splines and Tensor Product B-Splines

Given a non-decreasing sequence $\mathbf{u} = (u_0, u_1, \ldots, u_{p+1})$ we define a B-spline $B[\mathbf{u}]$: $\mathbb{R} \to \mathbb{R}$ of degree $p \geq 0$ recursively as follows [Sch81]

$$B[\mathbf{u}](u) := \frac{u - u_0}{u_p - u_0} B[u_0, \ldots, u_p](u) + \frac{u_{p+1} - u}{u_{p+1} - u_1} B[u_1, \ldots, u_{p+1}](u), \quad (2.1)$$

starting with

$$B[u_i, u_{i+1}](u) := \begin{cases} 1; & \text{if } u_i \leq u < u_{i+1}; \\ 0; & \text{otherwise,} \end{cases} \quad i = 0, \ldots, p.$$

We define $B[\mathbf{u}] \equiv 0$ if $u_{p+1} = u_0$ and terms with zero denominator are defined to be zero.

A univariate spline space can be defined by a polynomial degree p and a knot vector $\mathbf{u} = \{u_0, u_2, \ldots, u_{N+p}\}$, where the knots satisfy: $u_{i+1} \geq u_i, i = 0, \ldots, N + p - 1$, and $u_{i+p+1} > u_i, i = 0, \ldots, N - 1$, i.e, a knot value can be repeated $p + 1$ times. The number of times a knot value is repeated is called the multiplicity m of the knot value. The continuity of the spline function across a knot value of multiplicity m is C^{p-m}.

A basis for the univariate spline space above can be defined in many ways. The approach most often used is the univariate minimal support B-spline basis. In this, the B-splines are defined by selecting $p + 2$ consecutive knots from $\mathbf{u}$, starting from the first knot. So $B_{i,p}(u) := B[u_i, \ldots, u_{i+p+1}](u)$ is defined by the knots $u_i, \ldots, u_{i+p+1}$, $i = 0, \ldots, N - 1$. The minimal support B-spline basis has useful properties that ensure numeric stability such as local support, non-negativity and partition of unity (the basis functions sum up to one in all parameter values in the interval $[u_p, u_N]$).

Given two non-decreasing knot sequences $\mathbf{u} = \{u_0, u_1, \ldots, u_{N_1+p_1}\}$ and $\mathbf{v} = \{v_0, v_1, \ldots, v_{N_2+p_2}\}$ where $p_1 \geq 0$ and $p_2 \geq 0$. We define a bivariate TP B-spline $B_{i,j,p_1,p_2} : \mathbb{R}^2 \to \mathbb{R}$ from the two univariate B-splines $B_{i,p_1}(u)$ and $B_{j,p_2}(v)$ by

$$B_{i,j,p_1,p_2}(u, v) := B_{i,p_1}(u) B_{j,p_2}(v).$$

The support of B is given by the Cartesian product

$$supp(B_{i,j,p_1,p_2}) := [u_i, u_{i+p_1+1}] \times [v_j, v_{j+p_2+1}].$$

A bivariate TP spline space is made by the tensor product of two univariate spline spaces. Assuming that both univariate spline spaces have a minimal support B-spline basis, the minimal support basis for the TP B-spline space is constructed by making all tensor product combinations of the B-splines of the two bases. The minimal support basis for this spline space contains the TP B-splines $B_{i,j,p_1,p_2}(u, v)$, $i = 0, \ldots, N_1 - 1$, $j = 0, \ldots, N_2 - 1$. As in the univariate case, the basis has useful properties such as non-negativity and partition of unity.

2.3 Refinement of B-Spline Curves

Spline curves are frequently represented using a univariate minimal support B-spline basis.

$$f(u) = \sum_{i=0}^{N-1} c_i B_{i,p}(u), u \in [u_p, u_N].\qquad(2.2)$$

Here $c_i \in R^d$, $i = 0, \ldots, N - 1$ are the curve coefficients and d is the dimension of the geometry space. The curve lies in the convex hull of its coefficients.

A B-spline curve can be locally refined. Figure 2.1a shows a quadratic curve with knot vector $\{0, 0, 0, 1, 2, 3, 3, 3\}$. The curve coefficients and the control polygon corresponding to the curve are included in Fig. 2.1, and the associated B-splines are shown below. In Fig. 2.1b, a new knot with value 2 is added, thus increasing the knot multiplicity in an already existing knot. The curve is not altered, but the control polygon is enhanced with a new coefficient. The coefficient is marked with a circle in Fig. 2.1b. The double knot at 2 allows creating a curve with C^0 continuity. When moving the coefficient marked with a square, we obtain a sharp corner as in Fig. 2.1c. Note that only the last part of the curve is modified.

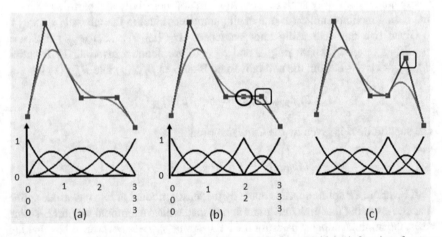

Fig. 2.1 Knot insertion into a quadratic B-spline curve. **a** Initial curve with basis functions, **b** curve and basis functions after knot insertion, **c** curve and basis functions after re-positioning of one coefficient

2.4 B-Spline Based Locally Refined Surface Methods

As described in Sect. 2.3, adding an extra local degree of freedom to a univariate B-spline basis just adds a knot, and an additional coefficient. However, adding an extra degree of freedom in one of the univariate minimal support B-spline basis of a TP B-spline represented surface adds an extra row or column of coefficients in the coefficient grid, with a resulting large increase in the bulk of the surface representation. Consequently, for application areas where it is desirable to add extra degrees of freedom/representation power locally just where needed, TP B-spline representations are not a good solution.

In some applications such as Isogeometric Analysis (IgA) [Hug05], and the representation of terrain and seabed, the lack of local refinement is a severe restriction. A TP B-spline surface covers a rectangular domain and the need for approximation power will not in the general case be uniformly distributed throughout the domain. In IgA the traditional shape functions local to each element in Finite Element Analysis are replaced by B-splines that cross element boundaries and connect elements with higher order continuity.

There are three main B-spline based approaches for extending spline surfaces to support local refinement [Dok19].

- **Hierarchical B-splines** (HB) [For88, Kra98] and **Truncated Hierarchical B-splines** (THB) [Gia12, Spl17] are based on a dyadic sequence of grids determined by scaled lattices. On each level of the dyadic grids a spline space spanned by uniform TP B-splines is defined. The refinement is performed one level at the time. For HB, TP B-splines on the coarser level are removed and B-splines at the finer level added in such a way that linear independence is guaranteed. The sequence of spline spaces for HB will be nested. To ensure partition of unity of the basis THB were introduced that truncate the TP B-splines of the hierarchical basis by TP B-splines from the finer levels.
- **T-splines** [Sed03, Sed04] denote a class of locally refined splines, most often presented using bidegree $(3, 3)$. In T-spline terminology control points are used rather than coefficients. Thus we stick to this terminology when addressing T-splines. The starting point for T-spline refinement is a TP B-spline surface with control points and meshlines (initial T-mesh) with assigned knot values. For bidegree $(3, 3)$ T-splines, the knot vectors of the TP B-spline corresponding to a control point are determined by moving in the T-mesh outwards from the control point in all four axis parallel parameter directions and picking in each direction knot values from the two first T-mesh lines and/or control points intersected. The mid knot value in each knot vector is copied from the start control point. The refinement is performed by successively adding new control points in-between two adjacent control points in the T-mesh. A new control point inherits one parameter value from the T-mesh line. The other parameter is chosen so that the sequence of control point parameter values have a monotone evolution along the T-mesh line. Control points on adjacent parallel T-mesh lines that have one shared parameter value are connected with a new T-mesh line. The general formulation of T-splines does not guarantee

a sequence of nested spline spaces. However, T-spline subtypes such as semi-standard T-splines and Analysis Suitable T-splines do so by imposing restrictions on how to refine, see Scott et al. [Sco11].

- **Locally Refined (LR) B-splines** [Dok13, Joh13] the refinement approach of this book, starts (as T-splines, HB and THB) from a TP B-spline surface. The refinement is performed successively by inserting axis parallel meshlines in the mesh of knotlines (from here on denoted the mesh). Each meshline inserted has to split the support of at at least one TP B-spline. The constant knot value of a meshline inserted is used for performing univariate knot insertion. The refinement is performed in the parameter direction orthogonal to the meshline in all TP B-splines that have a support split by the meshline. This approach ensures that the spline spaces produced are nested and that the polynomial space is spanned over all polynomial elements. In Sect. 2.5, we provide further details on additional refinements triggered and how the resulting TP B-splines can be scaled to achieve partition of unity.

Note that a main distinction between T-spline algorithms on the one side, and LR B-spline, HB and THB algorithms on the other side is that T-spline algorithms work in the mesh of control points and find the collection of B-splines traversing the mesh of control points, while LR B-spline, HB and THB algorithms directly refine the spline space and thus automatically ensure nested spline spaces.

2.5 LR B-Spline Refinement Method

The process of Locally Refined B-splines is described in detail in [Dok13]. Please consult the paper for formal proofs. Below a summary of the most important steps is presented.

The refinement always starts from a TP B-spline space $\mathcal{B}_0$. The refinement proceeds with a sequence of meshline insertions producing a series of collections of TP B-splines $\mathcal{B}_0, \mathcal{B}_1, \ldots, \mathcal{B}_k, \mathcal{B}_{k+1}$ spanning nested spline spaces each providing a refined surface

$$F(u, v) = \sum_{B \in \mathcal{B}_i} c_B s_B B(u, v), \quad i = 0, 1, \ldots, k, k + 1. \tag{2.3}$$

Above we end with $k + 1$, as we will now detail how to create $\mathcal{B}_{k+1}$ from $\mathcal{B}_k$.

Note that we require that all TP B-splines in these collections have minimal support. By this we mean that all meshlines that cross the support of a TP B-spline also have to be a line in the mesh representing the knots of the TP B-splines counting multiplicity. We denote these meshlines knotlines of the TP B-spline. Thus, to ensure that all TP B-splines have this property, the process of going from $\mathcal{B}_k$ to $\mathcal{B}_{k+1}$ frequently includes a number of intermediate steps and a corresponding sequence of intermediate LR B-spline collections.

Assume that Eq. 2.3 represents $F(u, v)$ using the collection $\mathcal{B}_k$ and we now want to represent $F(u, v)$ using a refined collection of minimal support TP B-splines $\mathcal{B}_{k+1}$. The refinement process goes as follows:

- As long as there is a $B \in \mathcal{B}_k$ that does not have minimal support on the refined mesh we proceed as follows. Let γ be a meshline that splits the support of B. This means that either γ is not a knotline of B, or γ is a knotline of B but has higher multiplicity than the knotline of B. Decompose B into its univariate component B-splines $B(u, v) = B(u)B(v)$. We now have two cases:

  - If γ is parallel to second parameter direction then it has a constant parameter value a in the first parameter direction. We insert a in the univariate B-spline $B(u)$ using Eq. 2.5 below and express $B(u)$ as $B(u) = \alpha_1 B_1(u) + \alpha_2 B_2(u)$. Then we make two new TP B-splines $B_1(u, v) = B_1(u)B(v)$ and $B_2(u, v) = B_2(u)B(v)$.
  - If γ is parallel to first parameter direction then it has a constant parameter value a in the second parameter direction. We insert a in the univariate B-spline $B(v)$ using Eq. 2.5 below and express $B(v)$ as $B(v) = \alpha_1 B_1(v) + \alpha_2 B_2(v)$. Then we make the two new TP B-splines $B_1(u, v) = B(u)B_1(v)$ and $B_2(u, v) = B(u)B_2(v)$.

B can be decomposed as follows, $B(u, v) = \alpha_1 B_1(u, v) + \alpha_2 B_2(u, v)$. We can now express $F(u, v)$ by replacing $B(u, v)$ with the two new TP B-splines. $F(u, v) = F(u, v) - c_B s_B B(u, v) + c_B s_B(\alpha_1 B_1(u, v) + \alpha_2 B_2(u, v))$. We update $\mathcal{B}_k$ by removing B and adding the TP B-splines $B_1(u, v)$ and $B_2(u, v)$. In addition we have to create/update both coefficients and scaling factors belonging to these two TP B-splines. We must have in mind that $B_1(u, v)$ and $B_2(u, v)$ often will be duplicates of B-splines already in $\mathcal{B}_j$. Now let $B_r, r = 1, 2$

- In the case B_r has no duplicate set $s_{B_r} = s_B \alpha_r$ and $c_{B_r} = c_B$.
- In the case B_r has a duplicate B_d set $s_{B_r} = s_{B_d} + s_B \alpha_r$ and $c_{B_r} = (s_{B_d} c_{B_d} + s_B \alpha_r c_B)/(s_{B_r})$, and remove the duplicate.

Note that s_B, α_r and s_{B_d} are all positive numbers, thus s_{B_r} will be positive.

When all $B \in \mathcal{B}_k$ have minimal support, we set $\mathcal{B}_{k+1} = \mathcal{B}_k$.

To simplify notation as in Chap. 3, we now define the scaled TP B-splines $N_B(u, v) = s_B B(u, v)$ to provide a basis that is partition of unity for the spline space spanned by $\mathcal{B}_j$. If $F(x, y) \equiv 1$ then all coefficients of the TP B-spline surface we start from, are 1. In this case the coefficients c_{B_r} calculated above all remain 1, duplicates or not. Consequently,

$$\sum_{B \in \mathcal{B}_k} N_B(u, v) = \sum_{B \in \mathcal{B}_k} s_B B(u, v) = \sum_{B \in \mathcal{B}_0} B(u, v) = 1. \qquad (2.4)$$

The process of inserting a knot $a \in (u_0, u_{p+1})$ into the local knot vector $[\mathbf{u}] = \{u_0, \ldots, u_{p+1}\}$ belonging to a univariate B-spline $B[\mathbf{u}]$, of degree p was first described by Boehm [Boe80]. We organize the resulting sequence of knots as a

Fig. 2.2 Parameter domain of an LR B-spline surface with indication on B-spline support. The mesh is shown as black lines. The support of two overlapping B-splines are shown in red and in blue

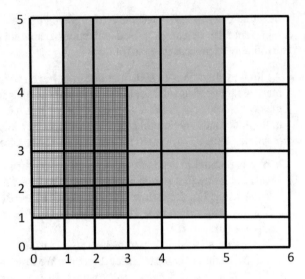

non-decreasing knot sequence $\{\hat{u}_0, \ldots, \hat{u}_{p+2}\}$. From this we make two new B-splines $B_1[\mathbf{u}_1]$ and $B_2[\mathbf{u}_2]$ with corresponding knot vectors $[\mathbf{u}_1] = \{\hat{u}_0, \ldots, \hat{u}_{p+1}\}$ and $[\mathbf{u}_2] = \{\hat{u}_1, \ldots, \hat{u}_{p+2}\}$. Then

$$B[\mathbf{u}] = \alpha_1 B_1[\mathbf{u}_1] + \alpha_2 B_2[\mathbf{u}_2], \tag{2.5}$$

where

$$
\begin{aligned}
\alpha_1 &:= \begin{cases} \frac{a-u_0}{u_p-u_0}, & u_0 < a < u_p, \\ 1, & u_p \le a < u_{p+1}, \end{cases} \\
\alpha_2 &:= \begin{cases} 1, & u_0 < a \le u_1, \\ \frac{u_{p+1}-a}{u_{p+1}-u_1}, & u_1 < a < u_{p+1}. \end{cases}
\end{aligned} \tag{2.6}
$$

The incremental refinement by knot insertion used by LR B-splines ensures that the spline spaces generated are nested. Figure 2.2 shows a parameter domain and the segmentation into boxes corresponding to a biquadratic LR B-spline surface. In addition the support of two TP B-splines is shown. We see that a meshline stops inside the blue TP B-spline support. This meshline is excluded from the local knot vectors defining the TP B-spline covering this support.

Linear independence of the resulting collection $\mathcal{B}_{k+1}$ of LR B-splines is not guaranteed, but occurrences are rare for the constructions used in this book. The approximation procedure outlined in Chap. 3 does not depend on linear independence. LR B-spline linear dependency configurations can, however, be resolved, if needed by an application, by insertion of extra meshlines. Linear dependence of LR B-splines is addressed in detail in [Pat20].

References

[Boe80]   Boehm, W. (1980). Inserting new knots into B-spline curves. *Computer Aided Geometric Design*. https://doi.org/10.1016/0010-4485(80)90154-2

[Dok13]   Dokken, T., Pettersen, K. F., & Lyche, T. (2013). Polynomial splines over locally refined boxpartitions. *Computer Aided Geometric Design*. https://doi.org/10.1016/j.cagd.2012.12.005

[Dok19]   Dokken, T., Skytt, V., & Barrowclough, O. (2019). Trivariate spline representations for computer aided design and additive manufacturing. *Computers and Mathematics with Application*. https://doi.org/10.1016/j.camwa.2018.08.017

[For88]   Forsey, D. R., & Bartels, R. H. (1988). Hierarchical B-spline refinement. In *SIGGRAPH 88 Conference Proceedings*, Vol. 4, pp. 205–212.

[Gia12]   Giannelli, C., Jüttler, B., & Speleers, H. (2012). THB-splines: The truncated basis for hierarchical splines. *Computer Aided Geometric Design*. https://doi.org/10.1016/j.cagd.2012.03.025

[Hug05]   Hughes, T. J. R., Cottrell, J. A., & Bazilevs, Y. (2004). Isogeometric analysis: CAD, Finite elements, NURBS, exact geometry, and mesh refinement. *Computer Methods in Applied Mechanics and Engineering*. https://doi.org/10.1016/j.cma.2004.10.008

[Joh13]   Johannessen, K. A., Kvamsdal, T., & Dokken, T. (2013). Isogeometric analysis using LR B-splines. *Computer Methods in Applied Mechanics and Engineering*. https://doi.org/10.1016/j.cma.2013.09.014

[Kra98]   Kraft, R. (1998). *Adaptive und linear unabhangige multilevel B-splines und ihre Anwendungen*, PhD thesis, University of Stuttgatt.

[Pat20]   Patrizi, F., & Dokken, T. (2020). Linear dependence of bivariate Minimal Support and Locally Refined B-splines over LR-meshes. *Computer Aided Geometric Design*. https://doi.org/10.1016/j.cagd.2019.101803

[Sch81]   Schumaker, L. L. (1981). *Spline Functions: Basis Theory*. New York: Wiley.

[Sco11]   Scott, M. A., Li, X., Sederberg, T. W., & Hughes, T. J. R. (2011). Local refinement of analysis-suitable T-splines. *Computer Methods in Applied Mechanics and Engineering*. https://doi.org/10.1016/j.cma.2011.11.022

[Sed03]   Sederberg, T. W., Zheng, J., Bakenov, A., & Nasri, A. (2003). T-splines and T-NURCCs. *ACM Transactions on Graphics*. https://doi.org/10.1145/882262.882295

[Sed04]   Sederberg, T. W., Cardon, D. L., Finnigan, G. T., North, N. S., Zheng, J., & Lyche, T. (2004). T-spline simplification and local refinement. *ACM Transactions on Graphics. doi, 10*(1145/1015706), 1015715.

[Spl17]   Speleers, H. (2017). Hierarchical spline spaces: Quasi-interpolants and local approximation estimates. *Advances in Computational Mathematics*. https://doi.org/10.1007/s10444-016-9483-y

Chapter 3
Adaptive Surface Fitting with Local Refinement: LR B-Spline Surfaces

Abstract A locally refined (LR) B-spline surface is a piecewise polynomial surface for which the distribution of the surface coefficients can be locally adapted. Such a mathematical representation is interesting for fitting scattered and noisy data, as the local behaviour of a real point cloud may require more degrees of freedom only locally. The number of redundant surface coefficients is minimized, which avoids the fitting of the point cloud's noise. The surface approximation is performed iteratively either by solving a least squares system or by a local approximation method. This procedure allows for mesh refinement in domains where the distance between a current surface and the point cloud exceeds a prescribed tolerance. In this way, parts of the LR B-spline surface obtained at previous steps may be kept unchanged. This chapter aims at explaining the adaptive fitting using local refinement with LR B-splines. We present two examples with simulated point clouds to illustrate the methodology.

Keywords LR B-splines surface · Adaptive refinement · Surface fitting · Local refinement · Least-squares · Multilevel B-spline Approximation (MBA)

3.1 Adaptive Local Refinement

In this chapter, we will let the x- and y-coordinates of the points serve as the parameter values while the surface (or function) approximates the z-component of the data. We note that the algorithm handles parameterized points as well, for which each point is given with a parameter pair and 3D coordinates, see [Flo05] for a review of the different parameterization methods. In the following we will focus on elevation and therefore denote the surface parameters x and y.

G. Kermarrec et al., *Optimal Surface Fitting of Point Clouds Using Local Refinement*,
SpringerBriefs in Earth System Sciences,
https://doi.org/10.1007/978-3-031-16954-0_3

3.1.1   General Principle of Adaptive Spline Refinement

We distinguish between two type of methods for fitting point clouds using local refinement:

- Adaptive methods: Here the spline space used in the surface fitting is only changed locally guided by the previous approximation result.
- Non-adaptive methods: the refinement of the spline space is performed independently of previous results.

Adaptive local refinement procedures are favorable in terms of the number of coefficients that defines the final surface. The proposed adaptive local refinement will be performed as follows:

1. Initial step: The point cloud is approximated by a tensor product (TP) B-spline surface which is a piecewise polynomial over a rectangular domain, see Chap. 2 for more details. A TP B-spline surface is a special case of an LR B-spline surface. This initial surface is represented as an LR B-spline surface to start the iteration.
2. Following steps: As long as the distance between some points and the surface is above the given tolerance in some subdomains of the surface, the surface is refined in these parts *only*.
   We address alternative refinement strategies in Sect. 3.2, while the technical details of the refinement procedure is described in details in Chap. 2 and briefly in Sect. 3.1.2. The principle of adaptive refinement is summarized in Algorithm 1.
3. Last step: A geospatial point cloud may represent a very rough terrain and contain noise and possible outliers. We usually stop the iteration before all points are closer to the surface than the tolerance to avoid fitting the noise. The refinement process can be stopped by the number of iterations or by statistical criteria as described in Chap. 4.

Algorithm 1: Principle of adaptive surface approximation with LR B-spline surfaces.

Data: Point cloud, maximum number of iterations, tolerance
Result: Approximating surface, information on approximation accuracy
Generate initial surface;
Compute accuracy;
while *there exist points with larger distance than the given tolerance and the maximum number of iterations is not reached* **do**
    Refine the surface in areas where the tolerance is not reached;
    Perform approximation in the current spline space;
    Compute accuracy;
end

3.1.2 Refining the LR B-Spline Surface

The initial surface is the starting point for the adaptive surface approximation. It is a TP B-spline surface represented as an LR B-spline surface. The parameter domain of such a surface is defined by a regular mesh of lines (the initial tensor mesh represented as a LR mesh). The meshlines split the domain into rectangles (boxes), each corresponding to a polynomial piece of the surface. Axis parallel meshline segments are successively inserted into the surface where the accuracy requirements are not met and have to satisfy:

- The meshline segment starts and ends at meshlines in the orthogonal parameter direction.
- The meshline segment has to split the support of at least one existing LR B-spline.

The resulting mesh defines a collection of none overlapping boxes. A box can touch but don't overlap adjacent boxes. The union of the boxes corresponds to the union of the boxes of the TP mesh. The B-splines spanning the spline spaces are created in parallel to the refinement of the LR mesh by splitting existing B-splines that no longer have minimal support. This way, the number of internal meshlines traversing the entire support of the B-spline does not exceed the polynomial degree. Note that meshline multiplicity has to be included in the count. Figure 3.1a shows an initial TP mesh corresponding to a biquadratic surface prepared for local refinement. The supports of two overlapping B-splines are shown with brown vertical lines and green diagonal lines. A new meshline (black) is entered. Since it covers the support of the brown B-spline, it is a legal refinement. In Fig. 3.1b, we see that the B-spline corresponding to the brown support is split. The new supports are depicted with brown vertical and brown horizontal lines, respectively. The line does not traverse the green support, and this B-spline remain unchanged. The new line has been included in the mesh, but not in the definition of the green B-spline. The line also leads to the

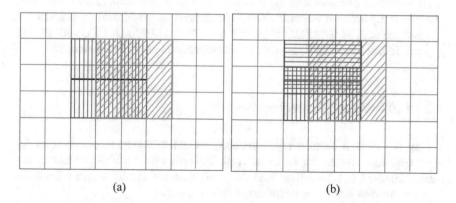

(a)   (b)

Fig. 3.1 Initial TP mesh (blue) with a meshline to insert (black). **a** The supports of two B-splines are shown prior to insertion. **b** After insertion, the brown B-spline is split while the green B-spline is unchanged. The black candidate line has become a part of the LR mesh and turned into blue

splitting of two more B-splines (not shown here for the sake of simplicity). Since two pairs of the new B-splines are identical, the total increase in the number of B-splines, and consequently the surface coefficients, is one.

3.1.3 Goodness of Fit of the Approximation

Following [Sky15], we define the performance indicators:

1. The root mean square error (RMSE) with respect to the approximated surface in the z-direction defined as $RMSE = \frac{1}{\sqrt{n_{obs}}} \|\hat{\mathbf{z}} - \mathbf{z}\|_2$, where $\hat{\mathbf{z}}$ is the estimated z-component of the point cloud obtained after the kth iteration. The RMSE is a good measure to estimate the standard deviation of a typical observed value from the model's prediction. Here we assume that the observed data can be decomposed into the predicted value and random noise with mean zero. The RMSE does not, however, take into account the spatial pattern of the error. Additionally, the number of observations will influence its value. A thinned version of a point cloud will give a smaller RMSE regardless of the accuracy in each single point.
2. The mean absolute error (MAE). To overcome the aforementioned limitation of the RMSE, we introduce the MAE, which is less sensitive to outliers than RMSE.

$$MAE = \frac{1}{n_{obs}} \sum_{i=1}^{n_{obs}} |z_j - \hat{z}_j|, \quad \mathbf{z} = \{z_j\}_{j=1}^{n_{obs}} \text{ and } \hat{\mathbf{z}} = \{\hat{z}_j\}_{j=1}^{n_{obs}}$$

We have $MAE \leq RMSE \leq \sqrt{n_{obs}} \times MAE$.
3. The maximum error defined as $Max_{err} = max \|\hat{\mathbf{z}} - \mathbf{z}\|$.
4. The number of points outside a given tolerance: n_{out}.
5. The number of coefficients n_{cp} estimated for a given iteration k of the refinement.
6. The computational time CT. The computations are performed on a stationary desktop with 64 GB of DDR4-2666 RAM. It has a i9-9900 K CPU with 8 cores and 16 threads, but a single core implementation is used in the examples.

3.2 Refinement Strategies

During the adaptive surface approximation, we maintain local information on the approximation accuracy for each mesh cell. The cells with insufficient accuracy are easily identified, but it is not obvious that every such cell should trigger refinement at every iteration step in the adaptive algorithm.

3.2.1 Isogeometric Analysis Versus Scattered Data Approximation

Some studies on the properties of various refinement strategies for LR B-splines and other LR splines exist, mostly in the context of Isogeometric Analysis (IgA). Johannessen et al. [Joh13] apply the strategies they named **Full span**, **Minimum span** and **Structured mesh** combined with LR B-splines for IgA. Bracco et al. [Bra18] also focus on IgA and study two classes of meshes for hierarchical B-splines. Hennig et al. [Hen17] compare two refinement strategies for hierarchical B-splines and T-splines. The refinement of LR B-spline surfaces from the perspective of maintaining local linear independence has been addressed in [Bre15] and [Pat20], respectively.

The use of LR splines for scattered data approximation differs from the use in IgA by the persistence of the surface. In the approximation setting, the computed surface is the final result while in IgA, the surface (or volume) is an intermediate step in the computations, i.e., the surface is not kept for further use. When approximating a point cloud, the variation of the underlying surface represented by the data points can be deduced during the approximation. A tailor-made selection of surface coefficients is possible. In IgA it is only possible to know which degrees of freedom are necessary first when the simulation is completed. An extensive introduction of new degrees of freedom may be appropriate. We will focus on approximation accuracy related to data size as in [Sky22].

3.2.2 Refinement Strategies in the Approximation Context

In the following, we focus on the approximation of geospatial data and present a variety of refinement strategies. The approximation algorithm coincides with the one presented in Sect. 3.1.1 and the iteration is pursued until the prescribed tolerance is met or no further accuracy improvements are possible. Here we give a short overview over the refinement strategies pursued in [Sky22], which are partly the same as in [Joh13]. The methods will be applied to various data sets in Chaps. 5 and 6.

The strategies are:

Full span (F): All B-splines overlapping a mesh cell identified for refinement are refined in their support. One new meshline for each selected parameter direction is entered in the middle of the knot interval(s) corresponding to the identified cell.

Minimum span (M): Only one B-spline overlapping a mesh cell identified for refinement is refined in its support. Several methods for selecting this B-spline can be addressed. In this context, we consider only a choice that combines the size of the B-spline support and the number of out-of-tolerance points situated in this support (c).

Structured mesh (S): A B-spline is selected for refinement, and refined in its support by inserting new knots in the middle of all knot intervals of the B-spline.

Restricted mesh (R):   A B-spline is selected for refinement, and refined in its support by inserting new knots in a subset of the knot intervals of the B-spline. Knot intervals towards the middle of the support, large intervals compared to the size of the support, and intervals corresponding to cells with a high number of points having a residual value larger than the tolerance are subject to knot insertion.

Given a mesh cell where the accuracy criteria are not met, the approximation flexibility will increase locally if one B-spline containing the cell in its domain is split. A good refinement strategy will insert enough new meshlines to resolve the criteria in a substantial number of cells without starting to adapt to noise. A slow pace in the introduction of meshlines in general gives surfaces with few coefficients, but increases the computational time to some extent. On the other hand, a very restrictive introduction of lines may lead to a blocking meaning that the approximation accuracy is not improved despite an introduction of new mesh lines in areas where the tolerance is not met. The pace of the refinement can be reduced by selecting a restrictive strategy (Mc or R), by refining in one parameter direction at a time (A) as opposed to refining in both directions (B), or by applying some kind of threshold. The latter means that not all candidate mesh cells or B-splines are refined at each iteration level. The candidates are sorted according to selected criteria involving the number of points outside the tolerance belt and the residual value in these points [Sky22]. Only the most significant candidate cells are refined.

To summarize, the full span strategy (F) shows the most stable behaviour for the diversity of test cases investigated. Alternating parameter directions (A) gives fewer coefficients than the full span strategy in general, with an acceptable increase in computational time. Thresholding can be beneficial, but these results were slightly less conclusive. Furthermore bidegree $(2, 2)$ is preferred over bidegrees $(1, 1)$ and $(3, 3)$. Bicubic polynomials generally produces more coefficients without delivering better accuracy. Bilinear polynomials often results in lean surfaces, but the stability suffers for some data sets and some refinement strategies. We will build on the results from [Sky22] and complement them with statistical considerations in Chap. 4. Further examples are given in Chaps. 5 and 6.

Figures 3.2 and 3.3 show the resulting mesh after refinement triggered by one mesh cell or B-spline using different strategies. The initial surface is biquadratic and defined on a uniform mesh with seven inner knots in each parameter direction. Normally, a number of refinements is performed at every iteration step, but here only one cell or B-spline support is selected for illustration purposes.

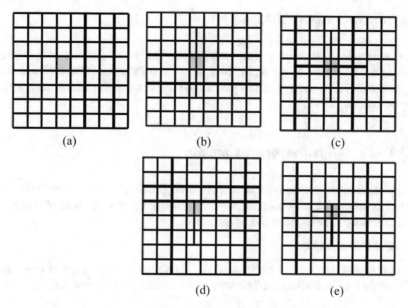

Fig. 3.2 Refinement strategies. **a** Initial mesh, selected cell is yellow. **b** Full span in one parameter direction at a time (FA). **c** Full span in both parameter directions (FB). **d** Minimum span in one parameter direction at a time (McA). **e** Minimum span in both parameter directions (McB)

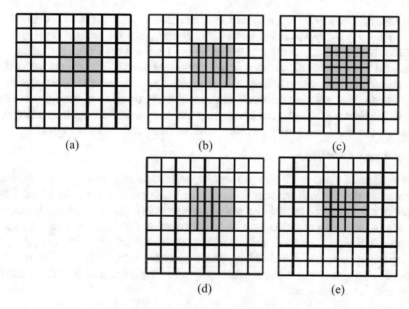

Fig. 3.3 Refinement strategies. **a** Initial mesh, selected B-spline support is yellow. **b** Structured mesh in one parameter direction at a time (SA). **c** Structured mesh in both parameter directions (SB). **d** Restricted mesh in one parameter direction at a time (RA). **e** Restricted mesh in both parameter directions (RB)

3.3 Surface Approximation

At each step in the iterative surface fitting algorithm, the coefficients of the current surface are computed to obtain the best fit to the point cloud. The approximation is performed either using a least squares (LS) approach or the multilevel B-spline approximation (MBA).

3.3.1 Least Squares Approximation

In this section, we introduce the concept of LS approximation within the framework of surface fitting. We restrict ourselves to the main formulas for the sake of simplicity. More details can be found in the references.

General Formulation

The LS approximation is a global method. The following expression is minimized with respect to the surface coefficients c_B over the entire surface domain:

$$\min_{\mathbf{c}}[\alpha_1 J(F(x, y)) + \alpha_2 \sum_{h=1}^{n_{obs}} (F(x_h, y_h) - z_h)^2]. \tag{3.1}$$

$\mathbf{x} = (x_h, y_h, z_h), h = 1, \ldots, n_{obs}$ are the data points. The surface is defined as $F(x, y) = \sum_{B \in \mathcal{B}_j} c_B N_B(x, y)$, where N_B, $B \in \mathcal{B}_k$, are the scaled TP B-splines defined as in Sect. 2.5. Note that we index by TP B-splines in the collection $\mathcal{B}_k$. Thus, the LS method minimizes a scaled version of RMSE. The expression is differentiated and turned into a linear, sparse equation system in the number of surface coefficients. The equation system is solved iteratively using a pre-conditioned conjugate gradient method. A pure LS approximation method will result in a singular equation system if there exist B-splines with no data points in their support.

Smoothness Function

A typical point cloud has a non-rectangular outline and may contain voids. Parts of the surface domain will frequently lie outside the domain of the point cloud. This challenge is addressed by adding a smoothness term $J(F(x, y))$ to the minimization functional. It allows to solve a non-singular equation system even if this situation occurs. Thus the smoothness term approximates the minimization of curvature and variations of curvature in the surface. These intrinsic measures are made parameter dependent to give rise to a linear equation system after differentiation. The smoothness term is expressed as:

$$J(F(x, y)) = \iint_{\Omega} \int_0^{\pi} \sum_{i=2}^{3} w_i \left(\frac{\partial^i F(x_0 + r \cos\phi, y_0 + r \sin\phi)}{\partial r^i} \bigg|_{r=0} \right) d\phi \, dx_0 dy_0. \tag{3.2}$$

At each point (x_0, y_0) in the surface domain Ω, a weighted sum of the directional first and second derivatives of the surface is integrated around the circle and the result is integrated over the surface domain. The directional derivative is represented by the polar coordinates ϕ and r. The two terms in $J(F(x, y))$ are given equal weight. The weight on the smoothness term is kept low to emphasize the approximation accuracy. In the examples of the next sections $\alpha_1 = 1.0 \times 10^{-9}$ and $\alpha_2 = 1 - \alpha_1$. A more detailed description of the procedure can be found in [Meh97]. A suite of alternative smoothness terms is presented in [Now98].

3.3.2 Multilevel B-Spline Approximation

MBA is a local, iterative approximation method, [Lee97]. It was originally developed for a hierarchy of TP B-spline functions. Given a current surface $F_k(x, y)$, the residuals corresponding to the data points are defined and a surface $G_k(x, y)$ approximating these residuals is computed. In the hierarchical B-spline setting, a set of subdomains where a prescribed tolerance is not met is identified and residual surfaces are computed for these domains only, [Zha98]. A special construction is applied to maintain continuity of the final hierarchical surface at the boundaries between these subdomains and the remaining parts of the surface.

Computing the Coefficients of the Residual Surface

The coefficients q_B of the residual surface $G_k(x, y) = \sum_{B \in \mathcal{B}} q_B N_B(x, y)$ are computed individually for each scaled TP B-spline N_B. Each coefficient q_B is determined from the collection of data points $P = (x_p, y_p, z_p) \in \mathcal{P}_B$ in the support of N_B. Let $r_p = z_p - F_k(x_p, y_p)$ be the residual corresponding to the point (x_p, y_p, z_p). The coefficient is computed as

$$q_B = \frac{\sum_{P \in \mathcal{P}_B} N_B(x_p, y_p)^2 \phi_{B,P}}{\sum_{P \in \mathcal{P}_B} N_B(x_p, y_p)^2}, \tag{3.3}$$

where $\phi_{B,P}$, addressed below, depends both on the residuals of the points in $\mathcal{P}_B$ and the collection, $\mathcal{B}_T$, of B-splines with a support overlapping at least one point in $\mathcal{P}_B$. By default $N_B \in \mathcal{B}_T$. For each of the residuals, we define an under-determined equation system

$$r_p = \sum_{C \in \mathcal{B}_T} \phi_{C,P} N_C(x_p, y_p), \quad P \in \mathcal{P}_B. \tag{3.4}$$

where $\phi_{C,P}$ are unknowns to be determined. There are many solutions to this equation system. For MBA, the solution is computed as a pseudo-inverse, see [Hsu92], giving

$$\phi_{C,P} = \frac{N_C(x_p, y_p) r_p}{\sum_{C' \in \mathcal{B}_T} N_{C'}(x_p, y_p)^2}, \quad C \in \mathcal{B}_T, \ P \in \mathcal{P}_B. \tag{3.5}$$

This solution minimizes

$$\sum_{P \in \mathcal{P}_B} \phi_{C,P}^2 \tag{3.6}$$

in the LS sense.

Finally we select $\phi_{B,P}$, $P \in \mathcal{P}_B$ as the missing piece in Eq. 3.3. The process is explained in more detail in [Zha98].

Updating the LR B-spline Surface

The current surface and the residual is represented by a collection of LR B-splines following the procedure of Sect. 2.5. Thus the updated surface can be computed as

$$F_{k+1}(x, y) = F_k(x, y) + G_k(x, y) = \sum_{B \in \mathcal{B}} (c_B + q_B) N_B(x, y). \tag{3.7}$$

MBA is an iterative process. Experience indicates that repeated applications of the approximation algorithm without adding more mesh lines improves the approximation accuracy. The improvement is stopped when the accuracy is restricted by the potential in the current collection of B-splines.

In [Sky15], the two approximation algorithms are compared in various examples. In general, the LS algorithm has a better approximation order while the MBA-algorithm is more stable when the spline space is less regular and/or the number of points in each element is low. We will discuss this topic further in Chap. 5.

3.3.3 Summary of the Adaptive Refinement

The LS method provides the best approximation to the point cloud in the L2 norm. Adding the smoothness term with a small weight maintains the good approximation properties. However, some supports of the smallest B-splines can contain none or few points in case of unevenly distributed scattered data points. Moreover, there is a risk of overfitting with LS, leading to ripples in the surface, see [Bra20]. Thus, the MBA method should be preferred after a few iterations with LS. As stated in Eq. 3.6, the pseudo inverse tends to minimize the deviation of the residual surface from zero. The procedure of local refinement combining LS and MBA is illustrated in Fig. 3.4 and can be summarized as follows:

1. The surface fitting algorithm starts with a LS approximation of the point cloud at the coarser levels to benefit from the best approximation property.
2. The local refinement strategy for LR-splines is applied iteratively to cells on which the L1-norm in the z-direction of the point-wise residuals exceeds a certain tolerance.
3. The algorithm switches to MBA after a few iterations. The switch is triggered either (i) by a prescribed iteration level being reached, (ii) if the convergence of the conjugate gradient method used for LS slows down, or (iii) if the number of

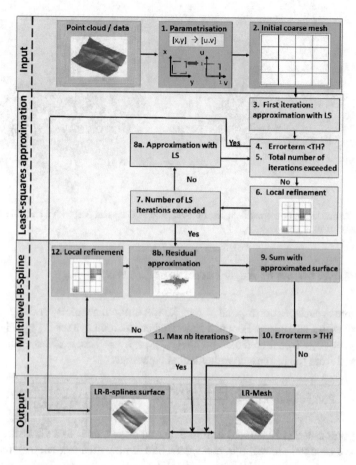

Fig. 3.4 Combination of MBA and LS for adaptive surface approximation with LR B-splines

surface coefficients and consequently the size of the LS equation system becomes large.

3.4 Example of Adaptive Refinement

In the following section, we will illustrate how adaptive fitting with local refinement using LR B-splines works. To that end, two simulated point clouds are generated.

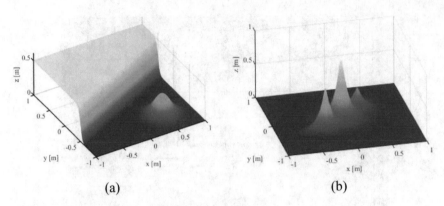

Fig. 3.5 Visualization of the mathematical functions. **a** A Gaussian bell with a dam-like jump. **b** Three peaks on a flat ground

3.4.1 Generation of Reference Point Clouds

The reference surfaces correspond to Fig. 3.5a: A smooth geometry and Fig. 3.5b: A geometry with sharp edges. For each generated point cloud, we set $(x, y) \in [-1, 1]^2$ with $n_{obs} = 40,000$ scattered data points (x_i, y_i, z_i), $i = 1...n_{obs}$. The z-component is obtained from the proposed mathematical equations:

$$\text{Point cloud (a): } z = \frac{\tanh(10y - 5x)}{4} + \frac{1}{5e^{(5x-2.5)^2 + (5y-2.5)^2}}. \tag{3.8}$$

The point cloud is illustrated in Fig. 3.5a and corresponds to a Gaussian bell as would be a mountain in real life, and a dam-like jump with a smooth transition.

$$\text{Point cloud (b): } z = \frac{1}{3e^{\sqrt{(10x-3)^2 + (10y-3)^2}}} + \frac{2}{3e^{\sqrt{(10x+3)^2 + (10y+3)^2}}}$$
$$+ \frac{3}{3e^{\sqrt{(10x)^2 + (10y)^2}}}. \tag{3.9}$$

The surface represents three peaks with different levels of altitudes on a flat ground, see Fig. 3.5b. We conjecture that the large gradient and the edges are challenging to approximate with LR B-splines. We refer to Chap. 1 for a discussion on approximation methods.

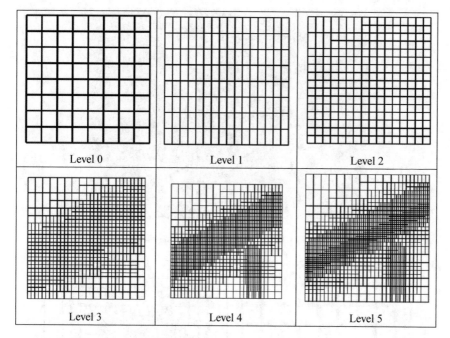

Fig. 3.6 Meshes at different level of approximation for point cloud (a). tolerance 0.007, refinement strategy FA, polynomial bidegree (2,2)

3.4.2 Results of Simulations

To illustrate how adaptive surface fitting with LR B-splines performs, we use the two point clouds (a) and (b). Following Algorithm 1, we set a maximum of 10 iterations and use a tolerance of 0.007. We present the meshes and some corresponding surfaces for point cloud (a) in Figs. 3.6 and 3.7, and in Fig. 3.8 for point cloud (b). Additionally, we compute the MAE, the maximum error, the number of points outside tolerance and the computational time at each iteration, see Tables 3.1 and 3.2 for point cloud (a) and (b), respectively. We focus on the full span refinement strategies with alternating parameter directions (FA), as described in Sect. 3.2 and choose the polynomial bidegree (2, 2) for the splines.

Point cloud (a)

After 4 iterations, the algorithm switches to the MBA strategy and stops after the 6th iteration. Here the number of points outside tolerance n_{out} reaches 0 and the maximum error Max_{err} is 0.0046. Approximately 1300 coefficients are needed to approximate the 40,000 points in less than 0.21 s. These values highlight the potential of surface fitting with LR B-splines to approximate in a short amount of time a high

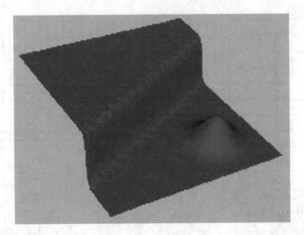

Fig. 3.7 Final fitted surface after 6 iterations for point cloud (a) with tolerance 0.007

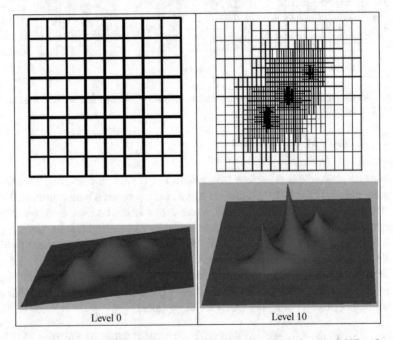

Level 0   Level 10

Fig. 3.8 Top: Meshes at different iteration steps for point cloud (b). tolerance 0.007, refinement strategy FA. Bottom: The fitted surface for level 0 and 10

Table 3.1 Results of fitting point cloud (a)

Level	CT (s)	n_{cp}	Max_{err}	n_{out}	MAE
0	0.028	100	0.174	31,831	0.0409
1	0.055	180	0.173	27,144	0.0283
2	0.084	310	0.087	12,176	0.0085
3	0.116	473	0.092	11,209	0.0079
4	0.151	751	0.167	7201	0.00871
5	0.185	1105	0.020	1137	0.00095
6	0.203	1336	0.0046	0	0.00041

0 corresponds to first iteration (coarser level). The refinement stops after 6 iterations as $n_{out} = 0$

number of observations. The MAE reaches 0.00041 at the last iteration level and is close to the tolerance after the 3rd iteration (0.0085 vs. 0.007). Here only 473 coefficients have to be estimated, which is 400 times lower than the total number of points. We point out that a lower tolerance would have led to a higher number of coefficients and more iterations. The choice of the tolerance is often left to the user's convenience and will be discussed in Chap. 4 by means of statistical criteria.

Figure 3.7 provides a visualization of the LR B-spline surface at the 6th iteration. This latter is close to the original one. At the 1st refinement level, the mesh is split by series of vertical meshlines. After the 3rd iteration the refinement is performed near the slope and the bell. As the number of iterations increases, the precision of the refinement in this area is getting higher.

Point cloud (b)

Unlike the approximation of point cloud (a), 4 points are still outside tolerance for fitting point cloud (b). They are located at the highest peak in the middle of the surface. We use the same tolerance of 0.007 and a maximum of 10 iterations. At that point, Max_{err} reaches 0.017 and the MAE 0.00037. Although the size of the two data sets is the same, the geometry of point cloud (b) is more challenging to approximate with an LR B-spline surface. We further note that MAE saturates at the 7th iterations. Increasing the levels of refinement from the 7th to the 10th does not lead to a significant improvement in accuracy (0.00039 vs. 0.00037) for 92 additional coefficients (800 vs. 892). We note that the CT remains under 0.3 sec after 10 iterations. This is a slightly higher CT than for point cloud (a) due to the increase of iteration levels. The splitting of each B-spline becomes more costly as the number of B-splines increases. From Fig. 3.8, the refinement leads to a mesh having more meshlines in the domain corresponding to the three peaks, which is expected from a local refinement strategy. The adaptive fitting with local refinement using LR B-splines has a high performance regarding the geometry with sharp edges and curvature changes (Table 3.2).

Table 3.2 Results of fitting point cloud (b)

Level	CT (s)	n_{cp}	Max_{err}	n_{out}	MAE
0	0.028	100	0.604	24,921	0.0219
1	0.054	180	0.505	10,811	0.0118
2	0.083	258	0.335	3925	0.0034
3	0.111	337	0.286	2536	0.0022
4	0.136	438	0.315	4484	0.0038
5	0.165	651	0.147	529	0.00076
6	0.182	754	0.076	113	0.00043
7	0.196	800	0.047	58	0.00039
8	0.213	856	0.016	31	0.00038
9	0.227	881	0.016	5	0.00037
10	0.237	892	0.017	4	0.00037

0 corresponds to first iteration (coarser level). The maximum number of iterations was set to 10. Tolerance 0.007

3.5 Conclusion

In this chapter, we have introduced the adaptive fitting with LR B-splines and explained four refinement strategies. The latter can be performed either in one or two directions; The choice of the method depends on the point clouds and the applications at hand, e.g., if the computational time or mesh regularity are important criteria. We have described the LS method to perform surface approximation, to which a smoothness term can be added. This approximation strategy is used in the first iterations of the adaptive algorithm, starting from a coarse mesh. In a second step, we described the MBA, which provides an explicit expression of the coefficients, thus avoiding matrix inversion without losing goodness of fit. An example with simulated observations have shown both the simplicity, conciseness and accurateness of the approximation method with local refinement.

References

[Bra18] Bracco, C., Giannelli, C., & VÁzquez, R. (2018). Refinement algorithms for adaptive isogeometric methods with hierarchical splines. *Axioms*. https://doi.org/10.3390/axioms7030043

[Bra20] Bracco, C., Giannelli, C., Großmann, D., Imperatore, S., Mokris, D., & Sestini, A. (2020). THB-spline approximations for turbine blade design with local B-spline approximations. ArXiv:2003.08706, https://doi.org/10.48550/arXiv.2003.08706

[Bre15] Bressan, A., & Jüttler, B. (2015). A hierarchical construction of LR meshes in 2D. *Computer Aided Geometric Design, 37*, 9–24.

[CCo] CloudCompare (version 2.12) [GPL software]. (2022). Retrieved from http://www.cloudcompare.org/

[Flo05] Floater, M. S., & Hormann, K. (2005). Surface parameterization: A tutorial and survey. In N. A. Dodgson, M. S. Floater, & M. A. Sabin (Eds.), *Advances in multiresolution for geometric modelling, mathematics and visualization*. Springer.

[Hen17] Hennig, P., Kästner, M., Morgenstern, P., & Peterseim, D. (2017). Adaptive mesh refinement strategies in isogeometric analysis–A computational comparison. *Computer Methods in Applied Mechanics and Engineering*. https://doi.org/10.1016/j.cma.2016.07.029

[Hsu92] Hsu, W. M., Hughes, J. F., & Kaufman, H. (1992). Direct manipulation of free-form deformations. *Computer Graphics (Proceedings of SIGGRAPH '92), 26*(2), 177–184.

[Joh13] Johannessen, K. A., Kvamsdal, T., & Dokken, T. (2013). Isogeometric analysis using LR B-splines. *Computer Methods in Applied Mechanics and Engineering*.

[Lee97] Lee, S., Wolberg, G., & Shin, S. Y. (1997). Scattered data interpolation with multilevel B-splines. *IEEE Transactions on Visualization and Computer Graphics, 3*(3), 229–244.

[Meh97] Mehlum, E., & Skytt, V. (1997). Surface editing. In M. Dæhlen & A. Tveito (Eds.), *Numerical methods and software tools in industrial mathematics* (pp. 381–396). Birkhäusser.

[Now98] Nowacki, H., Westgaard, G., & Heinemann, J. (1998). Creation of fair surfaces based on higher order smoothness measures with interpolation constraints. In H. Nowacki & P. D. Kaklis (Eds.), *Creating fair and shape-preserving curves and surfaces* (pp. 141–161). G. G. Teubner.

[Pat20] Patrizi, F., Manni, C., Pelosi, F., & Speleers, H. (2020). Adaptive refinement with locally linearly independent LR B-splines: Theory and applications. *Computer Methods in Applied Mechanics and Engineering*. https://doi.org/10.1016/j.cma.2020.113230

[Sky15] Skytt, V., Barrowclough, O., & Dokken, T. (2015). Locally refined spline surfaces for representation of terrain data. *Computers & Graphics*.

[Sky22] Skytt, V., & Dokken, T. (2022). Scattered data approximation by LR B-spline surfaces. A study on refinement strategies for efficient approximation. In C. Manni & H. Speleers (Eds.), *Geometric challenges in isogeometric analysis* (Vol. 49). Springer INdAM Series.

[Zha98] Zhang, W., Tang, Z., & Li, J. (1998). Adaptive hierarchical B-spline surface approximation of large-scale scattered data. In *Proceedings of Pacific Graphics 98. Sixt Pacific Conference*, pp. 8–16.

Chapter 4
A Statistical Criterion to Judge the Goodness of Fit of LR B-Splines Surface Approximation

Abstract The surface approximation obtained with adaptive strategies using locally refined (LR) B-splines depends on the degrees of freedom of the spline space, the tolerance from which the refinement is performed, the noise level of the scattered observations, the refinement strategy and the bidegree of the spline space. The choice of the best model is a challenging task that can be partially answered with statistical criteria, such as the Akaike Information Criterion (AIC). Here we relax the assumption that the approximation error should be normally distributed and with equal variance and propose the use of the student distribution to compute the AIC. We apply the AIC to decide which tolerance, refinement level, or polynomial bidegree are the most adequate for an optimal fitting. We highlight how the resulting AIC can be combined with more usual criteria to judge the goodness of fit of the surface approximation.

Keywords Information criterion · AIC · Surface approximation · t-distribution · Locally Refined B-splines · Local refinement

4.1 Introduction

A surface approximation of a point cloud can be done either globally (non-adaptive methods), or with locally adaptive methods. The adaptive surface fitting with LR B-splines used in this SpringerBrief belongs to the latter category. LR B-splines can be viewed as a generalization of univariate non-uniform B-splines, see [Dok13] for more details. The approximation of point clouds is performed step by step and the mathematical surface at each new iteration step depends on the result of the previous one [Sky22]. Contrary to Non Uniform Rational B-splines (NURBS) surfaces, local refinement is allowed. This method avoids overfitting in domains where no further refinement is necessary. We summarize the principle of adaptive surface fitting as follows:

- The starting point is a Tensor Product (TP) B-spline space. This is used for defining the initial LR mesh and the first collection of TP B-splines.

© The Author(s) 2023
G. Kermarrec et al., *Optimal Surface Fitting of Point Clouds Using Local Refinement*,
SpringerBriefs in Earth System Sciences,
https://doi.org/10.1007/978-3-031-16954-0_4

- The LR mesh is successively refined by inserting new meshlines: The first time in the initial LR mesh, later in the refined LR mesh. This insertion of meshlines is triggered from mesh cells, where at least one observation is associated with an error term higher than a given tolerance. The new meshline is extended to ensure that the support of at least one B-spline is completely traversed. The choice of the tolerance is linked with the level of accuracy needed, balanced by the computation time and number of surface coefficients to be estimated.
- The point cloud is approximated and the result is an LR B-spline surface.
- After a given number of iteration steps or as soon as no error term exceeds the tolerance, the final surface is computed.

The multilevel approximation (MBA) proposed in [Lee97] is often combined with a least-squares (LS) approximation in the first steps, see Chap. 3 for more details on the procedure.

The output surface depends on different parameters that are often chosen empirically. In this chapter, we will introduce a statistical criterion called the Akaike Information Criterion [Aka73] to judge the goodness of fit of the approximation, in addition to more usual values, such as the number of point outside tolerance or the mean absolute error (MAE). We further propose to investigate how the tolerance can be chosen with respect to the level of noise in the point clouds.

The remainder of the chapter is as follows: In the first section, we will describe the penalized model selection criteria within the context of surface approximation. The student or t-distribution will be introduced to face the challenge of outliers. Dedicated examples will show the potential of AIC as a global indicator for the goodness of fit.

4.2   Surface Approximation and Penalized Model Selection Criteria

To illustrate the challenges of choosing an optimal model in the sense of AIC, we will discuss two approaches: With and without a penalty term regarding the number of coefficients. We start with a data set of size n_{obs}, which we approximate with an LR B-spline surface by setting, e.g., the tolerance, the maximum number of iterations, the refinement strategy, and the polynomial bidegree of the spline. We call the result of the fitting *a model* and consider k possible models. The vector c_k contains the estimated coefficients and has a length n_{cp_k}. Each model has its own likelihood $L(c_k)$: This associates a numerical value to the question how "likely" the model is to the observations. It is convenient to work with the log-likelihood function for the model with the estimates c_k, which is defined as $l(c_k) = \log(L(c_k))$. The likelihood is a measure of goodness of fit and has a meaning *only* when it is compared with another likelihood computed for another model.

Approach 1 without penalty term: Case 1

When performing a surface approximation, one could search for the optimal refinement level, i.e., the iteration step from which the algorithm should be stopped because the optimal model has been found. Here we would call model 1 the approximation at level 1, model 2 at level 2. To each model is associated a likelihood, computed from the parameter vector of estimated coefficients. As the iteration step increases, its length will increase accordingly, but the corresponding likelihood may increase only slightly. Searching for the minimum of the likelihood without penalizing for the number of coefficients could lead to an overfitting and ripples in the approximated surface.

Approach 1 without penalty term: Case 2

If we make a first approximation of a scattered point cloud with a tolerance of 0.01, we obtain a parameter vector c_1 of length n_{cp1}; The approximation has a likelihood $L(c_1)$. In parallel, we can compute a second model by changing the tolerance to 0.005. Its likelihood is $L(c_2)$, with c_2 of length $n_{cp2} \gg n_{cp1}$. For both models, we stop the refinement after 5 iterations. Usually $L(c_1) \neq L(c_2)$ and we could state that $L(c_1) < L(c_2)$. This would lead to the conclusion that the second model is more appropriate to fit the data as its likelihood is higher. This statement is partially true: The number of coefficients for the second model is much higher than for the first one. This difference may be unfavorable (i) from a computational point of view, (ii) if overfitting should be avoided due to the presence of noise in the data, or (iii) if a lean model is preferred for storage or subsequent use. A too high number of coefficients should be avoided as ripples and oscillations may occur in the fitted surface.

The penalized criteria address the drawbacks raised in the first approach. In their simple form they are called the Bayesian Information Criterion (BIC) [Sch78] or the Akaike Information Criterion (AIC) [Aka73]. The two criteria are defined as:

$$BIC_k = -2l(c_k) + \log(n_{obs}) n_{cpk} \text{ and} \tag{4.1}$$

$$AIC_k = -2[l(c_k)] + 2n_{cpk}, \tag{4.2}$$

respectively. They can be seen as statistical alternatives to more usual heuristic considerations: The first term in Eqs. 4.1 and 4.2 is the log-likelihood, i.e., a measure of the goodness of fit to the data. The second term is a penalty term, which accounts for the increase in complexity. When k models are compared with each other, the model with the smallest IC is chosen. Choosing the best model within the framework of IC can be seen as finding a balance between these two quantities. The reader is referred to [Bur02] for the detailed derivation of the IC. In the following, we come back to the two cases with the second approach which accounts for a penalty term.

Approach 2 with penalty term: Case 1

For case 1, we can assume that the likelihood will saturate after a given number of iterations. At the same time, the number of coefficients will still strongly increase with each iteration step. It is likely that a minimum of the BIC and/or the AIC occurs, balancing both values.

Approach 2 with penalty term: Case 2

For case 2, only two models are compared with each other, the choice of the most optimal model is easy to meet if both $AIC_2 > AIC_1$ and $BIC_2 > BIC_1$, i.e., it can be concluded on the superiority of model 2 with respect to model 1: A tolerance of 0.005 is more optimal than 0.01 for approximating the data at hand, within the context of model selection with IC.

Potentially the BIC and the AIC may come to two different conclusions, i.e., $AIC_2 < AIC_1$ and $BIC_2 > BIC_1$. For case 1, this could be that the 3rd step is more optimal for BIC and the 5th for AIC. It is often stated that the BIC underestimates the optimal number of parameters to estimate. On the contrary, the assumption beyond the AIC is that the true model is unknown and unknowable. AIC is good for making asymptotically equivalent to cross-validation, and BIC for consistent estimation. In case of disagreement of the two criteria, other measures of goodness of fit should be added within the context of surface fitting, such as the MAE, the maximum error Max_{errk}, or n_{outk}.

In the following, we skip the subscript k for the sake of readability. We refer to Chap. 3 and recall that the following indicators to judge the goodness of fit will be used additionally:

1. The mean absolute error (MAE) defined as $MAE = \frac{1}{n_{obs}} \sum_{i=1}^{n_{obs}} |z_j - \hat{z}_j|$, $\mathbf{z} = \{z_j\}_{j=1}^{n_{obs}}$ and $\hat{\mathbf{z}} = \{\hat{z}_j\}_{j=1}^{n_{obs}}$. We have $\hat{\mathbf{z}}$ is the estimated z-component of the point cloud obtained after the kth iteration.
2. The maximum error is given by $Max_{err} = \max \|\hat{\mathbf{z}} - \mathbf{z}\|$,
3. The number of points outside a given tolerance: n_{out},
4. The degree of freedom or number of control points n_{cp} estimated for a given iteration step of the refinement,
5. The computational time CT. We have used a stationary desktop with 64 GB of DDR4-2666 RAM. It has a i9-9900 K CPU with 8 cores and 16 threads, but a single core implementation is used in the experiments.

These indicators are described in Chap. 3. Here we propose to highlight how they can be used in combination with the AIC to provide a weighted conclusion about the goodness of fit of the surface approximation.

4.3 Improving Information Criterion for Surface Approximation

We consider that AIC is an adequate criterion for model selection in the field of surface approximation as the true underlying surface is unknown. The risk of underestimation of the number of coefficients with the BIC should be avoided as details may not be revealed properly. If the AIC does not have a minimum, deeper investigations could be needed by changing the setup of the surface approximation (bidegree of the splines, refinement strategy, see Chap. 3). From now, we will only consider the AIC and search for its minimum when comparing k models.

4.3.1 The Challenge of Normality

The likelihood function is often taken to the Gaussian one, assuming the residuals of the surface approximation to be normally distributed. Unfortunately, this strong belief, when violated, can lead to a biased AIC. This compromises the correct and in-dubious determination of the AIC minimum and the choice of the most adequate model among a set of candidates. We propose to use the t-distribution (also called student's distribution), which gives more probability to observations in the tails of the distribution than the standard normal distribution [McN06]. This allows to give different weights to points outside the tolerance in the surface approximation, such as outliers. The t-distribution is defined by three parameters: μ its mean, σ its variance and v_t the degree of freedom of the distribution. The normal distribution is a special case of the t-distribution when the degree of freedom approaches infinity. The parameters of the t-distribution $\theta = [\mu, \sigma, v_t]$ cannot be expressed in a closed form. A stable approach to estimate them is via the iterative two-steps EM (Expectation Maximization) algorithm. In that case, we assume the observations to be independently and identically distributed. The so-called E-step computes the expected value of $l(\mathbf{p})$ given the observed data whereas the M-step consists of maximizing the expectation computed over the parameters to estimate. In most cases, the algorithm converges to a local maximum [Liu95].

The log-likelihood of the density of $\mathbf{r}$, with $\mathbf{r} = [r_1, ..., r_{n_{obs}}]^T$, the residuals of the surface approximation, which are assumed to come from the t-distribution $t(\mu, \sigma, v_t)$, is given by:

$$\log\left(L\left(r\right)\right) = n_{obs}\left(\log\Gamma\left(\frac{v_t+1}{2}\right)\right) - n_{obs}\log\Gamma\left(\frac{v_t}{2}\right) - \frac{1}{2}n_{obs}\log\sigma^2$$

$$+\frac{1}{2}n_{obs}v_t\log\left(v_t\right) - \frac{v_t+1}{2}\sum_{i=1}^{n_{obs}}\log\left(v_t + \delta\left(r_i; \mu, \sigma\right)\right) \qquad (4.3)$$

with Γ the Gamma function and $\delta\left(r_i; \mu, \sigma\right) = \frac{(r_i-\mu)^2}{\sigma}$, the standardized Mahalanobis square distance, refer to [Mah36] for more details.

The proposed AIC depends on the statistical properties of the approximation error of the fitted LR-surface: The size of the observation vector n_{obs}, the number of coefficients n_{cp}, the refinement strategy, the bidegree of the spline space, and the parameter of the t-distribution.

4.3.2   An Improved AIC for Surface Approximation

In real applications, the true model is unknown. It is easier to assess the potential of statistical criteria such as the AIC within the framework of simulations. We have chosen the two different point clouds presented in Chap. 3 to illustrate how the AIC can be used to judge the goodness of fit, as an alternative to the usual indicators (n_{cp}, Max_{err}, n_{out}, or MAE). We will investigate the optimal:

1. number of iterations,
2. refinement strategy,
3. bidegree of the spline,
4. tolerance with respect to the noise level.

4.4   The AIC to Choose the Settings for Surface Approximation of Scattered Data

The two reference surfaces used in the following correspond to (a): Smooth geometry and (b): Geometry with sharp edges. For each generated point cloud, we simulated $n_{obs} = 40,000$ scattered data points (x_i, y_i, z_i), $i = 1...n_{obs}$. Both are illustrated in Fig. 4.1. In the following, all values will be given in m, if not specified differently in the text. The z-component of point cloud (a) is given by

$$z = \frac{\tanh\left(10y - 5x\right)}{4} + \frac{1}{5e^{(5x-2.5)^2+(5y-2.5)^2}}. \qquad (4.4)$$

The point cloud (b) is generated by letting:

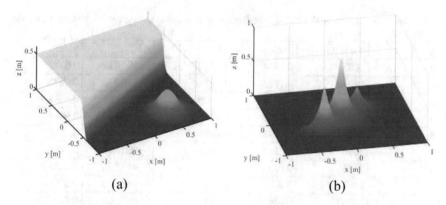

Fig. 4.1 Visualization of the generated point clouds. **a** A gaussian bell with a dam-like jump. **b** Three peaks on flat ground

$$z = \frac{1}{3e^{\sqrt{(10x-3)^2+(10y-3)^2}}} + \frac{2}{3e^{\sqrt{(10x+3)^2+(10y+3)^2}}} + \frac{3}{3e^{\sqrt{(10x)^2+(10y)^2}}}. \qquad (4.5)$$

To mimic real data, we add a Gaussian noise of standard deviation 0.002 m in the z-direction.

4.4.1 Number of Iteration Steps

Here we investigate the optimal level of refinement with the AIC. The results are presented in Table 4.1 for point cloud (a) and Table 4.2 for point cloud (b).

For point cloud (a), the number of points outside tolerance n_{out} was 0 after 15 iterations for a tolerance of 0.007 and a final MAE of 0.0016 m. The AIC has a local minimum at level 7 and a new minimum at level 11. The minimum at level 7 corresponds to the stage where the MAE saturates. This highlights the coherence of the different indicators for the smooth and homogeneous surface under consideration.

For point cloud (b), we found a minimum of the AIC at the 13th iteration level for the tolerance 0.007. Here there was no point outside tolerance after 14 iteration steps and the MAE reaches 0.0016. We note that at the 8th iteration, there is a turning point and both the MAE and the AIC saturate. Increasing the number of level of iterations leads to more coefficients (1650 at the 8th iteration step versus 2292 at the 14th) but not a strong improvement of the fitting. However, the improvement seems significant enough so that the AIC, which balances n_{cp} versus the likelihood, has a weak minimum at the 13th iteration step. We link this findings with the challenging geometry of the point cloud with peaks. The results are presented in Table 4.2 together with the other indicators for the sake of comparison.

Table 4.1 Investigation on the AIC by varying the iteration level for a given tolerance of 0.007 for point cloud (a)

Level	n_{cp}	Max_{err}	n_{out}	AIC	MAE
0	100	0.1797	30,765	− 185,812	0.0317
1	180	0.1777	25,805	− 198,156	0.0273
2	324	0.0946	11,095	− 306,991	0.0086
3	535	0.0937	10,612	− 310,462	0.0081
4	944	0.1638	9460	− 304,847	0.0111
5	1676	0.0247	1512	− 400,292	0.0022
6	2126	0.0154	242	− 418,056	0.0017
7	2467	0.0152	178	− 418,370	0.0017
8	2886	0.0168	136	− 418,208	0.0017
9	3247	0.0151	63	− 418,621	0.0017
10	3479	0.0103	38	− 419,037	0.0017
11	3619	0.0100	11	− 419,131	0.0016
12	3669	0.0088	10	− 419,091	0.0016
13	3712	0.0101	7	− 419,071	0.0016
14	3741	0.0081	2	− 419,127	0.0016
15	3752	0.0007	0	− 419,128	0.0016

The maximum Max_{err} and MAE are given in m. FA strategy is used for refinement, bidegree (2,2)

Table 4.2 Investigation on the AIC by varying the iteration level for a given tolerance of 0.007 for point cloud (b)

Level	n_{cp}	Max_{err}	n_{out}	AIC	MAE
0	100	0.6061	23,082	− 246,476	0.0205
1	180	0.5077	10,317	− 303,034	0.0121
2	291	0.6061	4193	− 369,168	0.0044
3	448	0.2994	2728	− 385,067	0.0033
4	628	0.2987	4899	− 363,989	0.0047
5	943	0.1483	722	− 411,889	0.0019
6	1198	0.0787	253	− 419,976	0.0017
7	1395	0.0521	194	− 420,539	0.0017
8	1650	0.0215	155	− 420,778	0.0016
9	1957	0.0197	79	− 421,697	0.0016
10	2206	0.0203	14	− 422,595	0.0016
11	2248	0.0172	7	− 422,613	0.0016
12	2271	0.0142	3	− 422,657	0.0016
13	2283	0.007	2	− 422,687	0.0016
14	2292	0.007	0	− 422,672	0.0016

The maximum Max_{err} and MAE are given in m, respectively. FA strategy is used for refinement, bidegree (2,2)

4.4.2 Refinement Strategy

In Chap. 3, we presented a set of refinement strategies that can be implemented with LR B-splines. We will here investigate two of them in the context of optimal surface fitting using AIC to judge the goodness of fit.

1. **FA** for which the refinement is performed alternatively in one of the two parameter directions,
2. **FB** for which the refinement occurs in both parameter directions at each iteration level.

The potential number of new coefficients at each iteration level is much less for FA compared to FB. For FA more iterations are expected to reach an acceptable accuracy. However, Skytt et al. [Sky15] show that this reduced pace in the introduction of new coefficients will lead to surfaces with fewer coefficients. Here the two refinement strategies can be considered as two models within the AIC framework as they are not equivalent, i.e., they lead to different residuals and likelihood. In the following, we set the tolerance to 0.007, the bidegree of the spline to (3,3) and the maximum iterations to 20. We compare two FA and FB refinement strategies to highlight the flexibility of the setting.

Point cloud (a)

We found that FB has a minimum AIC at the 7th iteration step but this latter starts to saturate at the turning point from which n_{cp} begins to increase strongly (4th iteration step), see Fig. 4.2. For FA, the n_{cp} increases at a slower pace compared to FB. The AIC has a weak minimum at the 15th iteration but saturates from the 6th one, as shown in Fig. 4.2.

The MAE for FA is 0.0016 after 20 iterations and a CT of 7.7 s. For FB, after 9 iterations and 3.8 s, the MAE reaches a comparable value of 0.00157. For both strategies, there is no point outside the tolerance at those iteration steps. Thus, FB is more favorable from a CT perspective. The computation times include computing the AIC. If AIC is omitted, the times are 1.10 and 0.95 s for FA and FB, respectively. However, the number of coefficients n_{cp} is much higher for the FB strategy (7024 vs. 4655 at the optimal iteration step). To compare, 4932 coefficients had to be estimated at the 4th iteration step with the FB strategy, and 173 points were still outside tolerance versus 2 points for the FA and 4655 coefficients.

We further note that the minimum of AIC for FB at the (from the AIC perspective) optimal 7th iteration is higher than for FA ($-$ 416,573 vs. $-$ 419,125 for FA at the optimal 15th iteration). This difference would indicate that FA is more optimal from a statistical criterion perspective than FB. This choice has to be weighted from a practitioner perspective, i.e., answering the question if more accuracy is needed or not, if the CT is an important criterion or not, and taking into consideration the challenge of overfitting. There is no definitive answer as the truth does not exist. It is a question of interpretation.

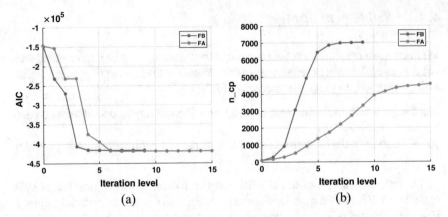

Fig. 4.2 Results of the approximation for the two refining strategies FA and FB, point cloud (a). **a** AIC. **b** n_{cp}

Point cloud (b)

For the point cloud (b), we find that FB has a minimum AIC at the 7th iteration step. It starts to saturate at the turning point (5th iteration step) from which n_{cp} begins to increase strongly, see Fig. 4.3. As the MAE, the AIC for FA and FB reaches a weak minimum, which indicates a fitting that can hardly be considered as optimal. We found that the MAE and the AIC for FA reach slightly lower values than for FB: for the MAE we found 0.0016 versus 0.0015 and for the AIC − 421,830 versus − 411,504 for FA and FB, respectively. The AIC can, thus, allow to conclude on the superiority of the FA strategy for point cloud (b) but the results differ slightly if we only consider the MAE as a criterion to judge the goodness of fit. This highlights the importance of accounting for n_{cp} to balance the likelihood. However the difference from a computational point of view for FA to reach the minimum is significantly higher than for FB: 1.437 s for FB versus 4.523 s for FA. The recording of the computational time includes computation of the AIC.

The previous results would tend to indicate that FA is more optimal than FB for fitting point clouds with LR B-spline surfaces. This is partially true and has to be weighted against CT. The the two examples clearly highlights that the fitting with the FB strategy produces more coefficients than FA for a similar accuracy while FA has a higher CT than FB. Still, this justifies our choice of using FA in the previous (and following) sections without lack of generality. This highlights, also, that new criterion should be found that also would also account for the CT to judge and balance the goodness of fit.

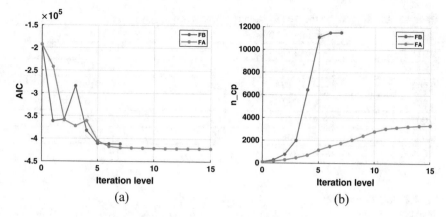

Fig. 4.3 Results of the approximation for the two strategies FA and FB, point cloud (b). **a** AIC. **b** n_{cp}

4.4.3   Tolerance

A proper tolerance is important for surface fitting: A large tolerance will make the process faster but may lead to underfitting, a smaller tolerance will increase the accuracy of the fitting result but costs more time, i.e., the fitting surface will be more complex, not to speak of the risk of overfitting. Hence, we can use AIC as the criterion to compare with the usual indicators and weight the number of parameters versus global accuracy. The fitting with minimum AIC is the optimal tolerance in a global sense.

In this section, we show the potential of AIC for investigating the tolerance. Here the standard deviation of the noise is taken as previously to be 0.002 m in the z direction. We vary the tolerance within a range from 0.005 to 0.011. We use refinement strategy FA and polynomial bidegree (2,2) and focus on point cloud (a). Similar conclusions could be drawn for point cloud (b) and are not presented here. Table 4.3 gives the AIC, as well as the iteration level with no point outside tolerance. For each tolerance, we set the number of maximum iteration steps to 20. For example, when the tolerance is 0.01, the approximation will continue until 14th iteration step, but the minimum AIC is reached at the 7th step. The AIC decreases with the tolerance and has a minimum for a tolerance of 0.007, which is illustrated in Fig. 4.4. This value was the optimal tolerance chosen for Table 4.1. We further note that the MAE stays around 0.0017 for all tolerances at the optimal number of iterations, and has a weak minimum at a tolerance of 0.006. This result is compatible with the results given by the AIC.

Table 4.3 Investigation on the AIC by varying the tolerance

Tolerance	Minimum AIC	Level	MAE
0.011	− 414,357	7	0.00183
0.01	− 415,525	7	0.00180
0.009	− 417,119	12	0.00174
0.008	− 418,927	19	0.00167
0.007	− 419,130	6	0.00169
0.006	− 418,508	10	0.00159
0.005	− 417,397	6	0.00162

Fig. 4.4 AIC with respect to tolerance

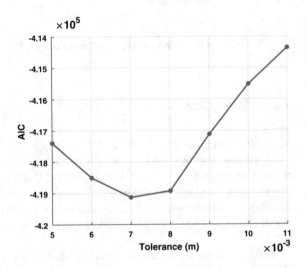

4.4.4   Polynomial Bidegree of the Splines

In this section, we vary the bidegree of the splines from (2, 2) (biquadratic) to (3, 3) (bicubic), which are usual choices for performing surface fitting. This corresponds to two different models within a model selection framework. We consider point cloud (a) and (b) and use the FA strategy for refinement, as well as a tolerance of 0.007. For point cloud (a) and for the optimal refinement level, we found that the biquadratic setting leads to a minimum of the AIC compared to the bicubic one (− 419,130 vs. − 419,125). From the MAE perspective, we found a value of 0.0016 for both settings at the optimal iteration step for the AIC (11th for the biquadratic and 15th for the bicubic respectively). The MAE does not decrease significantly for higher iteration steps, and, thus, does not allow to conclude in favour of a biquadratic or bicubic surface. Furthermore, a low MAE can be risky, i.e., linked with an overfitting. Here the AIC with its minimum, even if weak, has an evident advantage over the

MAE to find an optimal iteration level, by weighting the likelihood with the number of coefficients.

We have the same conclusion for point cloud (b). Here the minimum of the AIC is smaller for the bidegree (2, 2) (− 422,672 vs. − 421,830) but the MAE is similar for both optimal iteration steps corresponding to the minimum of the AIC (17 for the bicubic and 14 for the biquadratic).

Skytt et al. [Sky15] mentioned that in most cases a biquadratic surface will suffice, which is in accordance with our results. Thus, in most cases a higher bidegree of the polynomial doesn't contribute to a better accuracy of fitting LR B-spline surfaces for this type of data sets and noise levels.

4.4.5   Optimal Tolerance Versus Noise Level

Depending on the sensors and the conditions under which they are used, the noise level will vary. For a terrestrial laser scanner, the noise level of the range is known to depend on the intensity, i.e., the power of the backscattered laser signal recorded by the instrument after reflection. Atmospheric effects may also act as correlating the observations, i.e., decreasing the effective number of observations [Ker20]. The noise is often characterized by its standard deviation, a quantity which can be provided by the manufacturers. We can conjecture that a high noise level leads to a point cloud that is more challenging to fit optimally, with a strong risk of overfitting. Here we understand under overfitting "fitting the noise" instead of the true underlying surface. This effect is unwanted as it can give surfaces with ripples and oscillations [Bra20]. A wise choice of the tolerance can avoid or strongly mitigate the risk of overfitting. Thus the tolerance is an important parameter which is usually fixed rather empirically. Often, a low MAE is searched. Unfortunately, an artificially small error is not automatically linked with a high accuracy for fitting the underlying point cloud: In case of noise or outliers in the observations, even the contrary may happen.

We propose to investigate the choice of an optimal tolerance in the context of model selection, searching for a minimum of the AIC. To that end, we simulated different Gaussian noise vectors added to the reference point cloud. Their standard deviation was varied in a range of values between 0.001 m (low level of noise) and 0.0045 m. The noised surfaces were fitted with an LR B-spline surface. We chose the FA strategy and a biquadratic surface, following the results of the previous sections.

Here we vary the tolerance for a given noise level and search for the minimum AIC. Each AIC is computed at the optimal iteration step. We place ourselves in the framework of Monte Carlo simulations by simulating each time 100 noise vectors and taking the mean over all indicators.

The results of the investigations for point cloud (a) are presented in Fig. 4.5.

Figure 4.5 highlights that the optimal tolerance found with the AIC depends on the standard deviation of the noise level. As the noise level increases, the optimal tolerance increases, and so the AIC. We found a linear dependency of the optimal tolerance with respect to the noise level with a slope of 3 (left axis in Fig. 4.5a). This

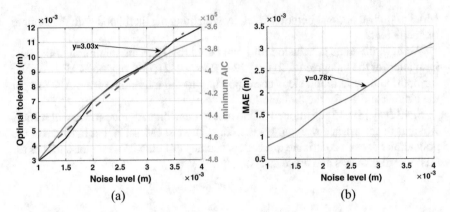

Fig. 4.5 Performance indicator versus noise level. **a** Optimal tolerance (left axis) and optimal AIC (right axis) versus noise level (std in m). **b** MAE (m) versus noise level

slope is slightly lower (close to 2) as the noise level increases. A similar result was found for point cloud (b) and is not presented here. The slope of 3 can be justified as corresponding to 3 times the standard deviation of the noise, i.e., this is the interval in which 68% of the measurements will fall assuming their normal distribution. We found that the number of optimal iteration steps stays between 6 and 7 and decreases as the noise level increases. This is an important finding as it is unnecessary -if not risky- to continue the adaptive refinement for noisy point clouds. This is what the AIC tells us. We further computed the MAE at the iteration step considered as optimal from the AIC, see Fig. 4.5b. We found a linear dependency, with a slope of 0.78. This latter is less predictable than the previous one regarding the noise level and will depend on the point cloud under consideration.

Following these results, we propose to choose the optimal tolerance as being 2.5 times the noise level. This is a good compromise when the noise of the sensor is unknown. Three times the noise level would be even more conservative and has to be weighted against a potential loss of accuracy.

4.5   Conclusion

In this chapter, we have introduced a statistical criterion as a new tool to judge the goodness of fit of the surface approximation. An information criterion is a weighted measure between the quality of fitting and the number of coefficients that are to be estimated. We showed how the AIC can come into play for determining the optimal level of refinement, the optimal bidegree of the spline, or the choice of the refinement strategy. Exemplary, we found that a biquadratic surface is optimal for a smooth point cloud. The tolerance is often fixed empirically with the aim to have a low RMSE or MAE. We found by investigating the AIC, that the optimal tolerance depends linearly on the noise level of the point cloud and can be fixed to 2.5 times the standard deviation of the noise of the observations. This information is often given by the manufacturers or can be guessed based on the residuals of the approximation and/or previous investigations. Thus, we have provided an answer to the question of the optimal tolerance with respect to the data at hand. These results will be used in Chaps. 5 and 6.

The use of the AIC to judge the goodness of fit is beneficial when many coefficients are needed to fit a point cloud: It avoids unnecessary steps and a possible overfitting. The AIC remains a global statistical quantity which has to be combined with other indicators, depending on what "optimality" should be for the application under consideration. For point clouds with high variability and local changes, the AIC only gives a global indication about the fitting.

References

[Aka73] Akaike, H. (1973). Information theory and an extension of the maximum likelihood principle. In B. N. Petrov & F. Csaki (Eds.), *Proceedings of the 2nd International Symposium on Information Theory* (pp. 267–281). Akademinai Kiado.

[Bra18] Bracco, C., Giannelli, C., Großmann, D., & Sestini, A. (2018). Adaptive fitting with THB-splines: Error analysis and industrial applications. *Computer Aided Geometric Design*.

[Bra20] Bracco, C., Giannelli, C., Großmann, D., Imperatore, S., Mokris, D., & Sestini, A. (2020). THB-spline approximations for turbine blade design with local B-spline approximations. *ArXiv:2003.08706*, https://doi.org/10.48550/arXiv.2003.08706

[Bur02] Burnham, K. P., & Anderson, D. A. (2002). *Model selection and multimodel inference*. Springer.

[Dok13] Dokken, T., Pettersen, K. F., & Lyche, T. (2013). Polynomial splines over locally refined box-partitions. *Computer Aided Geometric Design*.

[Dok19] Dokken, T., Skytt, V., & Barrowclough, O. (2019). Trivariate spline representations for computer aided design and additive manufacturing. *Computers & Mathematics with Applications, 78*, 2168–2182.

[Ker20] Kermarrec, G., Kargoll, B., & Alkhatib, H. (2020). Deformation analysis using B-spline surface with correlated terrestrial laser scanner observations: A bridge under load. *Remote Sens.*

[Lee97] Lee, S., Wolberg, G., & Shin, S. Y. (1997). Scattered data interpolation with multilevel B-splines. *IEEE Transactions on Visualization and Computer Graphics, 3*(3), 229–244.

[Liu95] Liu, C., & Rubin, D. B. (1995). ML estimation of the t distribution using EM and its extensions, ECM and ECME. *Statistica Sinica, 5*, 19–39.

[Mah36] Mahalanobis, P. C. (1936). On the generalised distance in statistics. *Proceedings of the National Institute of Sciences of India, 2*(1), 49–55.

[McN06] McNeil, A. J. (2006). *Multivariate t-distributions and their applications*. *JASA, 101*(473), 390–391.

[Sch78] Schwarz, G. (1978). Estimating the dimension of a model. *The Annals of Statistics, 6*, 461–646.

[Sky15] Skytt, V., Barrowclough, O., & Dokken, T. (2015). Locally refined spline surfaces for representation of terrain data. *Computers & Graphics*.

[Sky22] Skytt, V., & Dokken, T. (2022). Scattered data approximation by LR B-spline surfaces. A study on refinement strategies for efficient approximation. In C. Manni & H. Speleers (Eds.), *Geometric challenges in isogeometric analysis* (Vol. 49) Springer INdAM Series.

Chapter 5
LR B-Splines for Representation of Terrain and Seabed: Data Fusion, Outliers, and Voids

Abstract Performing surface approximation of geospatial point clouds with locally refined (LR) B-splines comes with several challenges: (i) Point clouds have varying data density, (ii) outliers should be eliminated without deleting features, (iii) voids, also called holes, or data gaps should be treated specifically to avoid the drop of the approximated surface in domains without points. These factors tend to be even more challenging when point clouds acquired from different sensors having different noise characteristics are fused together. The data set becomes non-uniform and the fusing process itself involves a risk of an increased noise level. In this chapter, we provide some tools to answer those specific challenges. We will use terrain and seabed data and show didactically how to perform adaptive surface approximation with local refinement and to select customized parameters. We will further address the problem of choosing an appropriate tolerance for performing an adaptive fitting, and discuss the refinement strategies within the context of LR B-splines. The latter is shown to provide a promising framework for surface fitting of heterogeneous point clouds from various sources.

Keywords Adaptive refinement · Surface fitting · Outliers · Voids · Trimming · Tolerance · Bathymetry · Data fusion · LR B-splines

5.1 Introduction

In this chapter, we will discuss a number of challenges that occur when working with geospatial point clouds, e.g., outliers, data fusion or domains without points. We will illustrate the discussion with a terrestrial LIDAR point cloud, or a seabed sonar point cloud and the combination of both. We aim to represent the area covered by both point clouds with a smooth parametric surface. The focus is on LR B-spline surfaces, which were shown to be appropriate within this context, see Skytt and Dokken [Sky22] for prominent examples.

The data acquisition of terrains and seabeds produces huge point clouds. The structure—or lack of structure—in the point clouds depends on the technology used to acquire them, i.e., on the sensor under consideration, may it be a terrestrial laser

© The Author(s) 2023
G. Kermarrec et al., *Optimal Surface Fitting of Point Clouds Using Local Refinement*,
SpringerBriefs in Earth System Sciences,
https://doi.org/10.1007/978-3-031-16954-0_5

scanner, a single or multibeam sonar. An efficient downstream use of the acquired data requires structured and compact data representations. Locally refined (LR) B-spline surfaces are smooth and flexible surfaces: They provide a middle road between the rigid but effective regularity of raster surfaces, and the highly flexible triangulated surfaces, see Chap. 1 for a detailed comparison. The LR B-spline surfaces are found to be convenient for representation of terrains and seabeds [Sky15, Sky16], as they accurately represent the smooth part of the data while having the flexibility to adapt to local shape variations without globally increasing the data size of the mathematical surfaces.

An LR B-spline surface belongs to the class of locally refined spline surfaces. It is a piecewise polynomial surface defined on a rectangular domain composed of axes parallel rectangular boxes (a mesh). In contrast to a tensor product (TP) spline surface, the boxes do not need to form a regular pattern. The concept of LR B-splines is described in detail in Chap. 2, which also includes an overview of alternative B-spline based locally refined surface methods.

The starting point for defining an LR B-spline surface is a TP B-spline surface. The adaptive refinement procedure inserts new meshlines into the surface description where the surface does not meet a prescribed accuracy requirement. The meshlines must satisfy the rule:

> A new meshline must split the support of at least one TP B-spline implying that the refinement increases the number of TP B-splines with at least one.

Refinement is performed in an iterative algorithm described in Chap. 3. First, the accuracy (L1 norm) of a current surface with respect to a given point cloud is computed. In a second step, the surface is refined where the accuracy does not meet the requirements. This process is repeated until the accuracy is found sufficient, which means that it does not exceed a predefined tolerance. Alternatively, the algorithm is stopped by some other constraint, normally the number of iterations. Figure 5.1 summarizes the surface approximation algorithm.

In the following, we propose to illustrate the fitting of a data set combining observations from terrains and seabeds. We will highlight how to solve different challenges that arise in real cases, such as

1. Outliers (outlier detection methods)
2. Data voids (bounding of coefficients and trimming)
3. Noise (selection of tolerance and method for surface approximation).

We will describe in detail the methods used to guide the user through methodological answers to real problems. More precisely, in a first section, we will present the data set that will be approximated. Next, outlier detection procedures will be compared. We will explain the concept of bounding the coefficients. Next the advantages of the Multilevel B-spline Approximation (MBA) will be highlighted. We will conclude by explaining the concept of trimming to deal with data voids. An appendix is dedicated to the output format of the LR B-spline surfaces.

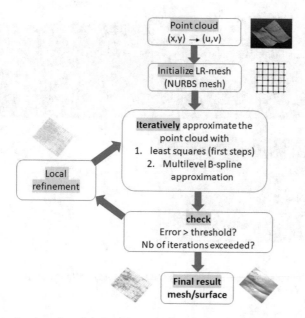

Fig. 5.1 Approximation of a point cloud by an LR B-spline surface with adaptive local surface refinement

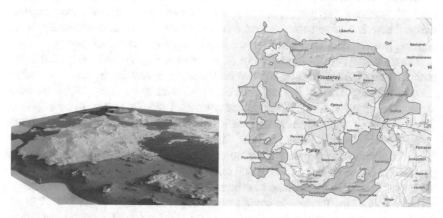

Fig. 5.2 The island Fjøløy in Norway

5.2 Description of the Data Set

Fjøløy is a 2.1 km^2 island in the municipality Stavanger at the south west coast of Norway, see Fig. 5.2. In this area terrestrial and bathymetry data are both available and represented in the same coordinate system. We select a set of corresponding land and sea data from the Fjøløy area, see Fig. 5.3. The terrestrial data set was obtained in 2016 with LIDAR. The bathymetry data was acquired in 2013 by a multibeam

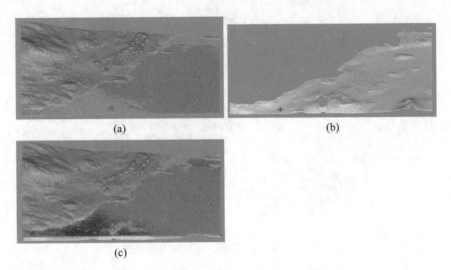

(a)   (b)

(c)

Fig. 5.3 Selected data sets from land (brown) and seabed (grey). **a** Terrestrial points, **b** bathymetry points, and **c** both

sonar and released for public use in the context of the project "Marine grunnkart pilot (Marine base map pilot)". The boat with the sonar has no access at very shallow water: There is a zone between land and sea where data is lacking (voids). In an ongoing project, data from this zone are acquired with LIDAR bathymetry.

The terrestrial data set shown in Fig. 5.3a, consists of 2,579,974 points containing both land and the sea surface. Several buildings, trees and stones can be found in the covered area. The data set has been classified prior to reception. 73 points were identified as outliers, 1,643,865 as ground points and 936,036 points are unclassified. The outliers are quite distinct. We consider this set as the reference to which we compare our outlier detection methods. The bathymetry data set, Fig. 5.3b, contains 25,107,199 unclassified points. Here no obvious outliers have been identified. The point cloud covers an area of 800×600 m, and the height range of the terrestrial data is $[-64.75, 156.32]$ m, including outliers. After the removal of outliers, the range is $[-0.74, 76.4]$ m. The height range of the bathymetry data is $[-21.13, -1.2]$ m. The total area covered by the combination of land and sea point clouds is just below 0.5 km^2.

We want to compute one LR B-spline surface combining both land and seabed as illustrated in Fig. 5.3c. To that aim:

1. We prepare the terrestrial data by removing the points from the sea surface.
2. We let the x- and y-values of the data points parametrize the surface, which consequently corresponds to a function representing elevation.
3. We eliminate outliers. This step is described in detail in the next section.
4. We use biquadratic polynomials for the surface approximation with LR B-splines. This choice balances the need for smoothness and flexibility.

5.3 Outlier Detection

An outlier is defined as an observation that lies at an abnormal distance from other values in a random sample from a population. Different strategies exist to eliminate them.

5.3.1 Strategies for Outlier Detection

Outliers in a data set can come from contaminated data samples, incorrect sampling methods, errors coming from the sensor during data collection or analysis [Haw80, Cha17]. Outliers in large geospatial data set can largely influence the results of the surface fitting, i.e., they will be adjusted as normal observations. If not excluded prior to the approximation, ripples or non-smooth surfaces are likely to arise.

The outliers are categorized as sparse outliers, or can come in clusters [Wan15], in which case they are more challenging to filter. Visual approaches are not suitable for large point clouds and automatic detection should be preferred. Some of the most popular methods for outlier detection for light detection and ranging (LIDAR) point clouds are reviewed in, e.g., Sotoodeh [Sot06]:

1. Z-Score or Extreme Value Analysis, as used in Sect. 5.3.2.2.
2. Linear Regression Models (Principal Component Analysis, Least Mean Square).
3. Probabilistic and statistical testing. Histogram, boxplot, Interquartile range (IQR Sect. 5.3.2.1) or Median Absolute Deviation are well known methods. From a statistical perspective, the Grubb's test can be used to identify single outliers as minimum or maximum value in a data set, or the Rosner's test for multiple outliers. The statistical tests are often limited to univariate data sets that follow approximately a normal distribution.
4. Clustering techniques. They are used to group similar data values into clusters having similar behaviour. Here it is assumed that outliers belong to any or only small clusters. Classification of ground points from a geospatial point cloud is related to outlier detection.
5. Deep learning based methods. We cite Pang [Pan21] for a review of different possibilities.
6. Surface or slope based methods. Roberts et al. [Rob19] test a set of surface-based and slope-based methods using some data sets known to be challenging to classify. With surface based methods, the ground surface is approximated iteratively. Ground points are identified using buffer zones defined from the parametric surface. Slope based methods assume that variations in terrain are gradual within a local neighbourhood.

5.3.2 Comparison of Outlier Detection Methods for the Selected Data Set

The terrestrial data set presented in Sect. 5.2 contains both outliers and unclassified points representing houses, trees, low vegetation and similar objects. Our aim is to identify outliers, but it is also interesting to see the extent to which outlier detection methods separate ground truth from vegetation and man-made objects. We will investigate the IQR and Z-score methods, as well as a method for detecting single outliers. The outlier detection is applied in the context of adaptive approximation of height data, see Chap. 3. The selected methods do not require any apriori estimate of the number of outliers. Moreover, they can be easily applied to a high number of data points.

The three methods are integrated in the adaptive approximation algorithm and applied at each iteration step in a regression setting. The distances between the points and the current approximating surface are compared making it a regression based method. The outlier detection is applied to subgroups of the data set identified by selecting the points situated in one mesh cell. The group testing is intended to reduce the computational effort in outlier detection. The point groups are subject for testing only if:

- The maximum distance between the subset of the point cloud and the surface is larger than a threshold, which depends on the maximum and the average distance between the surface and all data points in the previous iteration step.
- The local maximum distance has not be decreased significantly since the last iteration, which is an indication of the presence of at least one outlier.

The accuracy results from the last iteration step are obtained after outliers removal. At the start of the computation, it is hard to distinguish between features and outliers. As the surface is adapted to the point cloud, the distance in features will be smaller than for outliers; the criterion for allowing an outlier test is gradually decreased at each iteration step.

5.3.2.1 The IQR Test

Here the residuals correspond to a subset of the point cloud and the current surface are sorted according to their values. We call $Q1$ the first quartile of the residuals and $Q3$, the third one. Then $IQR = Q3 - Q1$ is the interquartile range of the residuals. We further denote two fences $f1 = Q1 - factor \times IQR$ and $f2 = Q3 + factor \times IQR$. An outlier is defined as a point with a residual value outside the range of these fences. Often $factor = 1.5$, justified by assuming that the residuals follow the normal distribution. This factor gives fences at $\mu - 3\sigma$ and $\mu + 3\sigma$, where μ is the mean and σ the standard deviation. This way 0.28% of the points are expected to be defined as outliers. Unfortunately there is no reason to believe that the residuals are normally distributed. A student distribution with a heavier tail is more probable,

Table 5.1 Outlier detection with IQR, various factors

Level	Max_{err}	MAE	n_{cell}	n_{test}	found$_{tot}$	Cell	n_{local}	$Max_{err,local}$	$f1$	$f2$	found$_{local}$
IQR factor 1.5											
1	100.45	0.902	64	32	9349	1	13,819	71.59	-4.23	3.94	13
						2	9288	93.99	-0.55	0.25	1,268
2	13.06	0.565	254	72	14,273	3	3355	164	-3.16	0.89	500
						4	3109	4.71	-6.133	7.092	0
3	6.60	0.334	827	169	5507	5	772	2.69	-1.22	0.91	55
						6	1273	5.34	-7.72	8.47	0
IQR factor 3											
1	100.45	0.902	64	32	1647	1	13,819	71.59	-7.29	7.01	1
						2	9288	93.99	-0.85	0.55	970
2	13.06	0.563	254	72	4540	3	3355	10.16	-4.68	2.41	369
						4	3109	4.72	-11.09	12.06	0
3	7.98	0.344	827	220	2202	5	738	2.61	-1.17	1.70	27
						6	1273	5.35	-13.83	14.59	0
IQR factor 5											
1	100.45	0.902	64	32	695	1	13,819	71.59	-11.38	11.10	1
						2	9288	93.99	-1.25	0.05	519
2	13.06	0.572	254	74	1395	3	3569	10.88	-7.58	6.09	71
						4	2944	4.90	-5.53	4.84	1
3	11.82	0.250	827	130	525	5	755	5.25	-2.43	2.08	26
						6	1396	4.45	-11.36	11.31	0

For each iteration step the maximum Max_{err} and average distance (MAE) is reported along with the total number of mesh cells n_{cell}, the number of cells tested for outliers n_{test} and the total number of identified (found) outliers found$_{tot}$. At each level, 2 example cells are selected reporting the number of points, fences and the number of identified outliers in the cell

see Chap. 4. As an assumption of the student distribution implies a more demanding computation to find the correct factor, we apply also the factors of 3 and 5 to our outlier detection and study the effect of this factor on the selected data set.

Table 5.1 shows some results for outlier detection with the IQR method. The total number of outliers identified is 29,129, 8,389 and 2,615 for an IQR factor of 1.5, 3 and 5, respectively. All obvious outliers are caught together with a certain amount of vegetation and house points, depending on the factor under consideration. The example cells at the first iteration step are the same for all factors, and we see that the number of outliers found are reduced with increasing factor. Similar distances between the subset of the point clouds and the surface lead to a very diverse number of outliers. Fortunately, this is not necessarily a problem: Large distances can also be synonymous with a low accuracy due to lack of freedom in the surface. This occurs typically at the beginning of the adaptive process when the steepness and roughness of the terrain varies in the selected area.

Figure 5.4 shows the result of the outlier classification. We see that much of the vegetation, buildings and some points at the sea surface are also classified as outliers for factor = 1.5, in addition to the obvious outliers (which are always found). In areas where the majority of the points belong to trees, see Fig. 5.4b, the position of the surface is influenced by the vegetation points as well as the ground points.

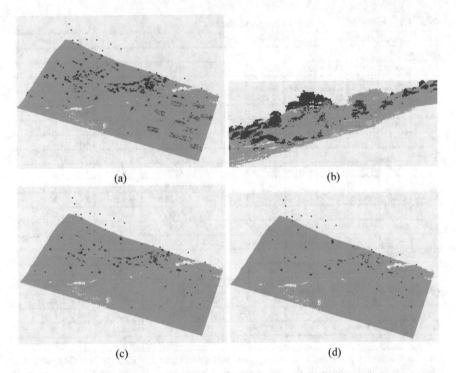

Fig. 5.4 The results of outlier detection with IQR, blue points are classified as outliers while the remaining points are light blue. **a** IQR factor 1.5, **b** detail with factor 1.5, **c** factor 3, **d** factor 5

Ground points and vegetation points become equally likely to be classified as outliers. When the IQR factor is increased, the part of the vegetation classified as outliers is decreased, but not eradicated, as illustrated in Fig. 5.4c and d.

The IQR outlier detection method removes many points and badly assumes that the data is normally distributed. Classification of points from vegetation and buildings should be done with more accurate methods. However, this method is simple and can give useful results if applied with care.

5.3.2.2   Z-score

Similarly to the IQR algorithm, the Z-score method is based on the assumption that the data are normally distributed. $Z_i = \frac{r_i - \mu}{\sigma}$ where r_i is residual number i, μ is the residual mean and σ the standard deviation. If the size of the residual is outside the range $[-3, 3]$, it is considered as an outlier. In theory, this should give the same result as the IQR test with factor 1.5 for a normally distributed data set. In this method the mean and standard deviation are computed explicitly; this puts less assumption on the distribution. Table 5.2 shows how outliers are detected with the Z-score method

Table 5.2 Outlier detection with the Z-score method

Level	Max_{err}	MAE	n_{cell}	n_{test}	std range	$found_{tot}$
1	100.445	0.90226	64	32	[0.81277, 1.6298]	428
2	13.055	0.57182	254	72	[0.15208, 3.2431]	3370
3	7.9887	0.34347	827	216	[0.13425, 3.4679]	1689

The global maximum and average distance between the surface and the points are reported along with the number of cells in the surface mesh and the number of cells tested for outliers. The standard deviation range applies to the tested cells and the number of outliers found the current iteration is reported in the last column

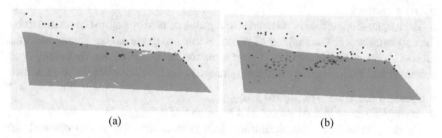

(a)   (b)

Fig. 5.5 Results of outlier detection with the Z-score method. Blue points are classified as outliers. **a** The result after one iteration, **b** the result after three iterations

during the adaptive surface approximation. The method classifies fewer points as outliers than the IQR method. The total number detected was 5487. In the first iteration step, only 1 or 2 points are found to be outliers in 27 of the cells tested. A point with a very large residual shadows for other points that may also be considered to be an outlier. These points can be found only when the most extreme cases are removed.

Figure 5.5 highlights that the evident outliers are detected in the first iteration step together with a few points corresponding to vegetation or buildings. In later iterations, more points close to the ground are added. Some points belonging to trees, bushes and houses are classified as outliers. Unfortunately, some points from a tree may be found to be outliers and some may not.

5.3.2.3 Detection Aimed at Single Outlier Points

The last outlier detection method to be investigated in this chapter is designed to fit within the context of adaptive surface approximation with local refinement such as, e.g., LR B-splines. It mainly aims at identifying single outlier points and has no direct link to the aforementioned statistical methods.

Table 5.3 Outlier detection aimed at single outliers

Level	Max_{err}	MAE	n_{cell}	n_{test}	Threshold	Found$_{tot}$
1	100.445	0.902264	64	32	25.788	34
2	13.055	0.571735	254	76	3.69255	14
3	11.8158	0.354466	827	139	3.2198	7

The global maximum and average distance between the point cloud and the surface are given along with the total number of cells, the number of cells where outlier detection is applied, the threshold for outlier detection and the found number of outliers

The points in a cell with a residual larger than the threshold are called candidate outlier points. The threshold is used in a pre-processing step to check the cells for possible outliers.

Each candidate outlier is compared to a group of nearby points not restricted by the cell boundaries. The number of points in this group varies, but should be close to 100. A set of characteristics is computed for the group of nearby points, both including and excluding the candidate outliers, to decide if they should be excluded:

- Standard deviation: std_{with} and $std_{without}$,
- Average distance to the surface: MAE_{with} and $MAE_{without}$,
- The range between their minimum and the maximum signed distance to the surface: R_{with} and $R_{without}$,
- Number of points: n_{with} and $n_{without}$.

For a candidate point to be classified as an outlier, the following rules must apply: $n_{with} - n_{without} \ll n_{with}$, $std_{with} \gg std_{without}$, $MAE_{with} \gg MAE_{without}$ and $R_{with} \gg R_{without}$. Furthermore, let z_o be the elevation of the candidate outlier point and z_p of the closest neighbouring points and r_o, and let r_p be the residual sizes for the two points. Then $|z_o - z_p| > 2 \times tol$ and $|r_o - r_p| > 2 \times tol$ where tol is the approximation tolerance. Moreover, a steep slope between the candidate outlier and the neighbouring point is required. The combination of these criteria implies that groups of outliers will be detected only if the group contains few points and/or is very deviant from other points in the neighbourhood.

Table 5.3 shows the number of points identified as outliers along with some additional information. We note that the number of outliers is much lower compared to the previous methods. After the most prominent outliers have been removed in the first step, the outlier threshold is reduced significantly. In the first step, the number of candidate outliers in the cell is one or two, and all candidates are classified as outliers. In the second and third step, the number of candidates in a cell varies from 1 to 101 and in most cases no outliers are detected. Groups of points belonging to houses, trees and other vegetation are tested and found not to be obvious outliers.

Figure 5.6 shows the location of the identified outliers. Mostly, the obvious cases are detected although a few points related to vegetation are included. As mentioned in the introduction, the data set contains 73 classified outliers, which were identified in a preprocessing step. The algorithm found 55 outliers where 49 also belong to the group of classified outliers. The current method is best adapted to the problem

(a)   (b)

Fig. 5.6 Results of outlier detection with the method aimed at single outliers. Blue points are classified as outliers. **a** The final result, **b** a detail

at hand (elimination of trees and vegetation), but it is complex and has a limited theoretical background. Nevertheless, it seems that a tailor made outlier detection is beneficial when combined with the adaptive method for surface generation.

5.4 Surface Approximation of the Selected Data Set

We use the points classified as ground from the terrestrial data set and remove the data points at the sea surface. This means that only points with a positive height component are included. The ground data is combined with the corresponding seabed data set resulting in the point cloud shown in Fig. 5.7. We notice that there are some shallow water areas where points are missing. Furthermore, the point cloud density is considerably higher for the seabed part compared to the terrain one: There are about 25 times more bathymetry points than terrestrial.

Fig. 5.7 Combination of terrain and sea data

5.4.1 Selection of Methods and Parameters

Figure 5.1 gives an overview of the surface approximation algorithm. The process starts from a TP B-spline surface, which is adaptively refined in areas where the distance between the surface and the point cloud is larger than a given tolerance. Given a current LR B-spline surface, we can perform the actual approximation with a least-square (LS) approach or multilevel B-spline approximation (MBA), see Sect. 3.3. LS approximation is a global approach with some best fit properties while MBA is an iterative explicit local approximation method. The method is to some extent expected to smooth out extreme behaviour in the approximating surface. We normally apply LS approximation for a number of iterations in the adaptive algorithm before turning to MBA.

Data sets are subject to noise and may contain outliers. It is, thus, not obvious that the approximation should be pursued until all points have a distance to the surface smaller than a given tolerance. Normally, the process is stopped by a maximum number of iteration steps, but the finding the optimal number of iterations is challenging. Computing the minimum of AIC is an alternative to find this optimum, but the process is time consuming. Moreover, it is a global method that does not take local variations in the point cloud into account, i.e., a minimum does not always exist. A tolerance is applied to identify where the surface needs to be refined. This value should be defined depending on the measurement accuracy, information that is not always known. Also, the actual selection of new meshlines to insert influences the accuracy and number of coefficients in the final surface. Various refinement strategies are discussed in [Sky22] and a short resume is given in Sect. 3.2. In the remainder of this section, we will discuss the selection of methods (MBA, LS, combination of both, refinement strategy) and parameters (tolerance, number of iterations) for the selected data set.

5.4.1.1 LS Approximation Versus MBA

Figures 5.8 and 5.9 compare (i) the approximation with LS until about 33,000 coefficients are estimated, and switch to MBA, and (ii) MBA for the entire computation using a tolerance of 0.5 m. We see that for the LS approximation, both the number of unresolved points and the average distance in these points are lower than MBA for the same number of coefficients. The difference is the largest for few coefficients and diminishes when the number of coefficients increases.

Figure 5.10 shows the approximating surfaces using LS and MBA. The difference is small, but looking at Fig. 5.11, it is clear that MBA offers a smoother transition in areas with no point. LS approximation should be applied early in the approximation process for data sets with relatively uniform density, a low noise level and no outliers. For non-smooth data sets with voids, MBA should be the preferred choice.

When fitting point clouds with spline surfaces, an overshoot in areas with steep gradients may arise, in particular with unevenly distributed data points. On the other hand, the surface is bounded by its coefficients due to the property partition of unity.

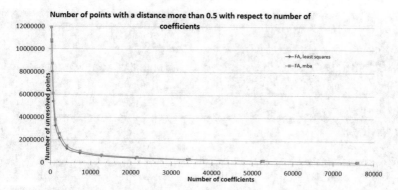

Fig. 5.8 Number of unresolved points with respect to the number of surface coefficients

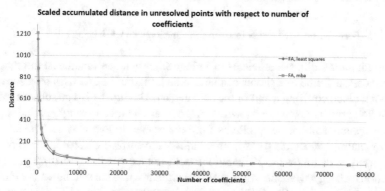

Fig. 5.9 Accumulated distance in points with distance more than 0.5 m scaled with a factor of 1/10,000

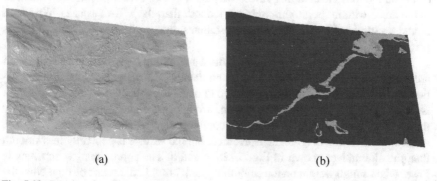

Fig. 5.10 Results of surface generation, LS = green, MBA = brown. **a** The surfaces are roughly similar, **b** the point set is included in the figure to emphasize the areas without points

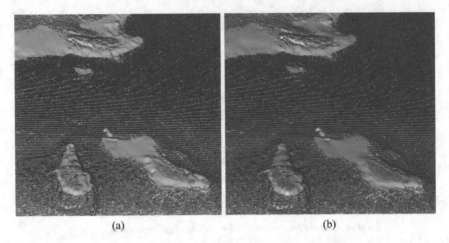

(a)   (b)

Fig. 5.11 Focus on areas without points. **a** Approximation with LS, **b** using MBA

By limiting the surface coefficients to a range slightly larger that the height range of the data set, extreme overshoots can be avoided. Figure 5.12 focuses on a subset of the point cloud covering a part of the area depicted in Fig. 5.11. The subset contains 1,494,242 points, which are shown in Fig. 5.12a. The area is rough, includes parts without points and has steep climbs from the seabed to two islands. The adaptive approximation procedure starts from a biquadratic surface without inner knots and is allowed to continue for 12 iterations with refinement in alternating parameter directions. Using MBA, the result is almost similar when the size of the coefficients are bounded or not, see Fig. 5.12b and c. This is not the case for the LS approximation. When the coefficients are bounded to a range slightly larger than the elevation range of the data point, the resulting surface depicted in Fig. 5.12d is quite well behaved in the areas without point although less smooth than the MBA surfaces. Without a bound on the coefficients, the surface oscillates drastically in areas without points (Fig. 5.12e).

The LS approximation is combined with a smoothing term to ensure a solution in areas without points, see Chap. 3 for more details. The weights on the approximation term and the smoothing term sum up to one. Normally, the weight on this term is kept low to emphasize approximation. We apply a higher weight (0.1) to study the effect of smoothing in challenging configurations as shown in Fig. 5.12f. The extreme behaviour in Fig. 5.12e is avoided, but the surface is generally less smooth than the alternatives shown in Fig. 5.12b, c and d. The approximation accuracy is lower when a high weight on the smoothing term is applied. It must be noted that the approximation errors increase in the last iteration step in theses cases. Otherwise, the accuracy does not differ much between the various approaches, see Table 5.4. LS approximation may become less accurate when the LR mesh gets very unstructured. Then the algorithm switches to perform approximation with MBA. We stop the iteration just before this situation occurs so the results are achieved with LS

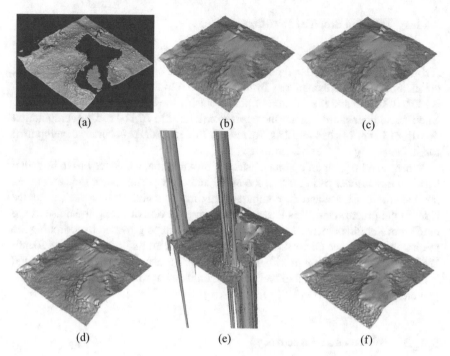

Fig. 5.12 Focus on areas without points. Approximation with different selections of approximation method. **a** Data points, **b** approximation with MBA and bound on the coefficients, the surface is light blue and the points can be glimpsed in clear blue, **c** approximation with MBA and no coefficient bounds, **d** LS approximation with coefficient bounds, **e** LS approximation, no coefficient bounds, **f** LS approximation with high weight on the smoothing term (0.1) and no coefficient bounds

Table 5.4 Accuracy of the subset of the point cloud with LS approximation and MBA

Method	Max_{err}	MAE	n_{out}	n_{out} (%)	n_{cp}
LS, bounds	2.388	0.113	32,260	2.16	3255
LS, bounds[a]	3.228	0.145	57,764	3.87	3238
LS, no bounds	2.388	0.113	32,194	2.15	3255
LS, no bounds[a]	5.454	0.146	58,869	3.94	3280
MBA, bounds	2.292	0.116	37,702	2.52	3203
MBA, no bounds	2.292	0.116	37,699	2.52	3203

The weight on the smoothing term for LS is $1.0e^{-9}$ and 0.1. The tolerance is 0.5 m
[a] Smoothing term has weight 0.1

approximation or MBA, purely. We note that the bounds on the surface coefficients do not hamper the approximation accuracy.

5.4.1.2   When to Stop the Iteration

Figures 5.8 and 5.9 indicate that the gain in continuing the approximation after the surface having 20,000–30,000 coefficients is small. For approximation with MBA, the maximum distance decreases from 3.782 to 3.426 m, the average distance from 0.100 to 0.073 m and the fraction of points outside the tolerance from 1.9 to 0.57% when the number of coefficients increases from 21,572 to 76,110 and the computation time from 3 min. 24 s to 4 min 28 s. We refer to Table 5.5 for the accuracy development for an increasing number of iterations.

When searching for an optimal surface approximation, a balance has to be found between the number of iterations, the MAE and other performance indicators, i.e., the maximum distance and the computational time for a given tolerance. The choice is let to the practitioner: This latter should judge the risk of fitting the noise as the number of iterations increases. Here an indication can be provided by searching the minimum of AIC, see Chap. 4. In our particular case, no minimum could be found: We link the lack of minimum with the fact that the surface contains many details and is not smooth enough, i.e., a global criterion on its own is not sufficient to judge the goodness of fit.

5.4.1.3   Tolerance and Accuracy

A main concern regarding surface fitting is linked with the accuracy of the approximation. This is especially important in areas like seabed shallows, while the noise level may be high at shallows due to sea vegetation and a narrow sonar width resulting in multiple traversals by the boat carrying the sonar. The surface should accurately represent the main shape of the terrain, but not necessarily adapt to every little stone. The tolerance is used to determine where the surface needs refinement and consequently the achievable accuracy. It is a predetermined value that should reflect the precision of the measurement. A level of 2–3 times the measurement error can be considered appropriate as discussed in Chap. 4. This is a first indication as the real error is normally larger than the precision of the measurement device, which is not always known. Several scans are merged and arbitrary objects, like power lines and fishes, may influence the result. Here we investigate the impact of the threshold on the fitting.

The surface approximations in Figs. 5.8, 5.9, 5.10 and 5.11 were performed with a tolerance of 0.5 m. The algorithm was allowed to run for 12 iterations, and all mesh cells where the maximum distance between the surface and a point in that cell exceeded the tolerance triggered refinement. All B-splines with the cell in its support were refined in one parameter direction at the time, in the x-direction at odd levels and the y-direction at even levels. This corresponds to the refinement strategy called FA, see Chap. 3 for more details. The MAE dropped below the tolerance at iteration level 2 for both LS approximation and MBA, and touched 0.1 m at level 9. The tolerance of 0.5 m is selected somewhat arbitrary, but is found to balance surface size and accuracy.

Table 5.5 Tolerance, number of iteration steps, MAE, number of coefficients and percentages of points with a distance to the surface in specified ranges

Tol	Steps	MAE	n_{cp}	$|d| < 0.4$ (%)	$0.4 < |d| < 0.6$ (%)	$0.6 < |d| < 1$ (%)	$|d| > 1$ (%)
0.1	12	0.069	217,002	98.77	0.93	0.25	0.05
0.1	9	0.1	32,212	96.62	2.28	0.94	0.17
0.1	6	0.156	4322	91.20	5.12	2.76	0.92
0.4	12	0.071	104,171	98.77	0.94	0.25	0.05
0.4	9	0.1	25,565	96.62	2.28	0.94	0.17
0.4	6	0.156	4257	91.20	5.12	2.76	0.92
0.5	**12**	**0.073**	**76,097**	**98.74**	**0.97**	**0.25**	**0.05**
0.5	9	0.1	21,567	96.61	2.28	0.94	0.17
0.5	6	0.157	4086	91.19	5.12	2.76	0.92
0.6	12	0.077	57,564	98.64	1.06	0.25	0.05
0.6	9	0.1	19,042	96.59	2.30	0.94	0.17
0.6	6	0.157	3911	91.18	5.13	2.77	0.92

The applied refinement strategy is FA and surface approximation with MBA is applied. Distances are given in m, and $|d|$ denotes the absolute value of the distance between a point and the surface

Table 5.5 presents some accuracy results for a selection of tolerances and maximum iteration levels. The setup used in Figs. 5.8, 5.9, 5.10 and 5.11 is highlighted with bold font. The difference in accuracy between the applied tolerances is remarkably small while the numbers of coefficients differ greatly when a high number of iterations is applied. In the first iteration steps, the selected tolerance plays a limited role. The approximation error indicates similar refinements for all applied tolerances.

Figure 5.13 shows that the configuration of points with a residual value smaller than or larger than 0.4 m is relatively similar for the tolerances 0.1 and 0.6 m. Some differences can be spotted mainly due to an increase in point size for the points with a

(a) (b)

Fig. 5.13 Point cloud coloured according to the distance to the surface. White points are closer than 0.4 m, green points lie below the surface and red points above. More saturated colour means larger distance. The size of the white points are reduced compared to the coloured points. **a** Tolerance 0.1 m, 12 iterations, **b** tolerance 0.6 m, 12 iterations

distance larger than 0.4 m in the picture. The surface adapting to a tolerance of 0.1 m
has more points within this tolerance belt than the other surfaces, but the difference
is negligible compared to the difference in the number of surface coefficients. The
percentages of points within this small belt after 12 iterations is 79.3, 78.4, 77.4
and 75.8% for tolerances of 0.1, 0.4, 0.5 and 0.6, respectively. The roughness of the
data does not allow such a tight approximation with a smooth surface. The majority
of the points with a high distance to the surface belong to the seabed. This can be
caused by the number of bathymetry points being much higher than terrestrial points,
but also from the bathymetry points being unclassified whereas terrestrial points are
classified as ground. The descent is most prominent in shallow seabed areas.

5.4.1.4 Refinement Strategies

In Sect. 5.4.1.3, we saw that a tighter tolerance increased the number of surface
coefficients considerably at later iteration levels without improving the accuracy
significantly. The effect of the extra refinement is low. Similar results were also
found in [Sky22]. A rapid introduction of new meshlines leads to more coefficients for
similar accuracy, but also a lower computational time. A slower pace in introducing
new degrees of freedom often led to few coefficients and an acceptable computation
time, while a very restrictive introduction could block further accuracy improvements
and eventually lead to more surface coefficients that contribute little to an accurate
approximation.

Table 5.6 illustrates how different refinement strategies for defining new meshlines
influence the approximation results. We stop the iteration after the surface has reached
20,000 coefficients. The number of iterations required is reported in column two. For
the strategies whose name starts with F and Mc, the refinement is triggered by mesh
cells that contain points with a residual value larger than the tolerance. For strategies
starting with S and R, refinements are triggered for B-splines having such points
in their support. If the strategy is marked by "all", all such occurrences will lead
to refinement while "tn" indicates that only mesh cells or B-spline supports with
a relatively high number of out-of-tolerance points combined with a large distance
to the surface will trigger refinement. Strategies marked with B will refine in both
parameter directions in each iteration step while strategies marked with A will refine
in alternating parameter directions. Strategies starting with F are full span strategies
meaning that all B-splines having the identified cell in its domain are split. Mc are
minimum span strategies. Here, only the one B-spline is defined to be refined and the
criterion is a combination of size and number of associated out-of-tolerance points.
For S strategies the identified B-spline is refined in all knot spans, while for R the
knot spans containing most out-of-tolerance points are refined. McA tn is the most
and SB the least restrictive refinement strategy in the list. We refer to Chap. 3 for
more details on each refinement strategy.

In Table 5.6, we compare the results for the different strategies after the last
iteration. We see that the "A" strategies always need a higher computational time
than the "B" since the refinement in "A" is performed in each direction separately

Table 5.6 Refinement strategies and associated accuracy results

Strategy	level	Max_{err}	MAE	n_{out}	n_{cp}	n_{in}/n_{cp}	CT
FA all	9	3.78	0.1	498,528	21,567	**1185.99**	3 m 16 s
FA tn	11	3.49	0.089	278,282	22,024	1171.38	4 m 12 s
FB all	5	3.41	0.088	356,038	39,019	659.18	2 m 19 s
FB tn	6	3.42	0.082	189,155	35,025	739.12	2 m 44 s
McA all	10	3.41	0.09	351,166	27,159	947.22	3 m 52 s
McA tn	12	3.41	0.088	221,103	23,882	1082.64	5 m 4 s
McB all	5	3.4	0.09	361,776	30,157	852.7	2 m 24 s
McB tn	6	3.38	0.089	229,529	25,848	999.97	2 m 58 s
SA all	9	3.78	0.099	498,113	26,558	963.12	3 m 13 s
SB all	5	3.41	0.087	355,637	51,124	503.11	2 m 16 s
RA all	10	3.41	0.092	372,451	27,996	923.6	3 m 53 s
RA tn	12	3.42	0.088	219,545	23,683	1091.8	4 m 53 s
RB all	5	3.41	0.093	384,317	32,251	796.64	2 m 23 s
RB tn	6	3.36	0.085	210,713	30,666	843.47	2 m 55 s

The iteration is stopped after 20,000 surface coefficients (n_{cp}) is reached. Distances are reported in m and computational time in min and s. The approximation efficiency is computed as number of points with a distance less than the tolerance (n_{in}) divided by the number of coefficients (n_{cp}). Thus, a high efficiency number is beneficial. The tolerance is 0.5 m and the number of data points is 26,076,683

implying that the number of coefficients to estimate is higher. However, the final number of coefficients relative to the accuracy tends to be lower for the "A" methods and the efficiency is higher. For this data set, SB has the lowest computational time but a high number of coefficients and the poorest efficiency among the recorded strategies. The best approximation efficiency is found for FA (marked with bold font). However, the efficiency does not take the value of the residuals into account as long as it is smaller than the prescribed tolerance; here the actual distance could be considered as well. We see that some methods will have lower computational time than FA. Thus, if the time is regarded as more important than the number of surface coefficients, FB and McB are good alternatives, preferably with some restrictions on the mesh cells that trigger refinement (tn). The results in this experiment fall well in line with the conclusions in [Sky22]. The choice of the refinement strategy could be also seen as a model selection problem, following the concept described in Chap. 4.

5.4.2 Dealing with Missing Points and Voids: Trimming

In Computer Aided Design (CAD), trimming is used to remove extra lines or extra parts of an object, see, e.g., Marussig and Hugues [Mar18] for an overview of methods in Isogeometric Analysis (IgA). Trimming aims to optimize the modeling and visualization of the approximated surface. Here we apply it to handle data gaps and "cut" the domains where no points were available for fitting. Often, this would have led to unfavorable ripples or voids as the algorithm tries to approximate without data support. We note that the parameterization and the mathematical description of the surface remain unchanged after trimming. We summarized the principle of trimming as follows:

1. We bound the points by curves in the xy-plane. These curves are often B-splines curves or NURBS.
2. The curves are arranged in one loop for the outer boundary and one loop for each hole and associated to the parameter domain (xy-plane for points parameterized by their x- and y-values) of the surface.
3. The outer loop is counter clockwise oriented, while eventual inner loops are clockwise oriented. By convention only the areas of the surface situated to the left of such trimming loops are considered valid. Consecutively, the loops divide the resulting trimmed patch into distinct parts where the direction of the curves tells which parts of the domains are visible or not.

Figure 5.14 explains the computation of trimming loops and the trimmed surface. A polygon of horizontal and vertical lines in the xy-plane surrounding the points is

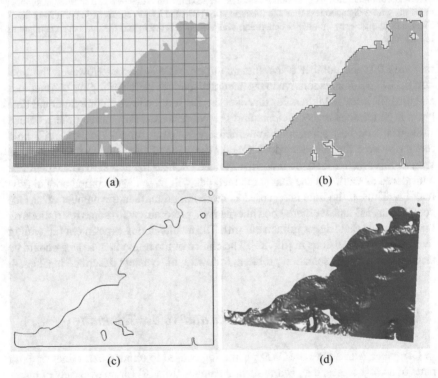

(a)   (b)

(c)   (d)

Fig. 5.14 Computation of the trimmed surface. **a** The point cloud (in khaki green) is recursively divided into subsets of the point cloud and bounded by polygons, **b** the composite polygons bounding the entire point cloud, **c** the polygons are approximated by a set of spline curves, **d** the final trimmed surface

computed in a recursive procedure. Depending on the density of the point cloud, a maximum recursion level is selected. A dense point cloud allows more recursions and consequently a more accurate polygon. The point cloud is recursively divided into blocks as shown in Fig. 5.14a. Here the maximum recursion level is two. The boundary lines of the blocks containing points are collected, while removing lines that occur twice. This happens when two adjacent blocks contain points. The resulting lines are sorted to create one or more polygons, see Fig. 5.14b. In Fig. 5.14c, the polygons are divided into pieces, each being approximated by a spline curve, and finally, in Fig. 5.14d, the trimmed surface is shown.

5.5 Conclusion

Adaptive LR B-spline surface approximation is a flexible method to "transform data into information". Within a context of approximating geospatial data, huge, noisy and scattered data set from terrains or seabeds can be represented in a compact way. The surface approximation with LR B-splines has following advantages:

1. The computational time is manageable.
2. The data storage is strongly simplified: Millions of points are condensed into a manageable number of coefficients to estimate.
3. The adaptive approximation method is flexible. The MBA can be combined with the LS approximation. Here the LS method is used in the first iterations, and the smoothness term can be adapted to avoid fitting of noise. In the last iterations, the MBA allows an explicit yet very accurate fitting. Because it has similarities with the L1 noise, outliers and data gaps can be optimally handled to keep the approximation smooth. This property is often needed for geospatial data set.
4. The refinement methods can be adapted depending on the data at hand (point density, presence of noise or outliers). Different parameters such as the tolerance, the polynomial degrees of the spline surface or the refinement strategies can be chosen individually.
5. The fit of the approximation can be judged using simple statistical concepts such as the mean absolute distance, the number of points outside tolerance or the maximum error. Additional statistical quantities, such as information criterion can provide orientation for optimizing the surface approximation.
6. The format is flexible and allows an export as TP B-spline surface in usual GIS software.
7. The C++ functions are freely available to permit a wide usage of the LR B-spline surface approximation, up to individual adaptation of the algorithms.

In this chapter, we have highlighted these properties and approximated a data set composed of seabed and terrain data recorded from sensors having different noise properties. More specifically:

1. We have compared different pre-processing strategies to eliminate outliers, and found that the method identifying single outlier points with no direct link to

statistical methods hits the target best: it reduced the risk of eliminating features that need to be approximated but found real outliers.

2. We have developed the concept of adaptive approximation, starting from a coarse mesh. A refinement is performed in cells where the error between the mathematical surface and the points exceeds a predefined tolerance.

3. We have highlighted how to deal with data voids that are a common challenge for many GIS data set. Here the point density may be so low that no plausible surface approximation can be performed. We have highlighted that MBA performs well in such cases. It is a computational advantageous method as no minimization has to be done.

4. We compared different parameters set up to achieve the best goodness of fit, e.g., the tolerance, the number of maximum iterations, or the refinement strategy. We have investigated different refinement strategies and shown that the FA (full span refinement in one direction at each iteration) was more favorable. We further showed how the tolerance affects the noise fitting.

5. We explained how a trimming can be performed to cut domains without points for which the fitting is unfavorable (ripples, oscillations).

The result of the surface approximation with LR B-splines is a mathematical surface with few coefficients in comparison to the huge number of points to approximate. The surface describes the underlying ground with high accuracy, which can be assessed by means of simple statistical quantities. Ongoing research tries to find the most optimal surface with respect to the data at hand by setting, e.g., the tolerance less empirically. To that aim, concepts developed in Chap. 4 can be used for smooth and homogeneous point clouds. In Chap. 6, we will present further applications of the LR B-spline surface approximation, such as deformation analysis with LR B-spline volume, or the drawing of contour lines from the mathematical model.

1   Appendix: Output Format and Source Code

An LR B-spline surface is stored in an ASCII file using doubles for the storage of coefficients. It is also supported by Part 42 of ISO 10303 (the STEP standard). The export of the LR B-splines surfaces to other formats is crucial for further processing of the mathematical surfaces. Exemplary, raster is the standard representation for terrains and seabed in current GIS software. Unfortunately, this representation does not support the same level of detail as an LR B-spline surface of the same area in general. To circumvent that challenge, different possibilities exist:

1. We compute a highly accurate LR B-spline surface, which gives rise to rasters of different resolutions. Thus, the LR B-spline surface can serve as a master representation to be harvested according to needs.

2. We extend the LR B-spline surface to a TP B-spline surface. Here the main drawback is a potential large increase in data size. Furthermore, this conversion contradicts the idea of LR splines.

3. We export the LR B-spline surface as a set of Bezier surfaces alternatively.

4. A better option is to represent the LR B-spline surface by a collection of TP B-spline surfaces maintaining the feature of data size distributed according to needs. To that aim, the LR B-spline surface can be divided into TP B-spline surfaces by the means of dedicated knot line insertions. The division into TP B-surfaces is performed by a recursive algorithm. This division is also an ingredient in the computation of contour curves and some details are given in Chap. 6, Appendix 1.

Please note that for all computation, we made use of the GoTools library module LR Splines 2D. The source code is freely made available by SINTEF Digital, Department of Mathematics and Cybernetics for downloading at the link: https://github.com/SINTEF-Geometry/GoTools/wiki/Module-LRSplines2D. The hardware requirements are Windows, Linux or MacOS. The program language is C++. Following software are required: Cmake, Boost, and Qt for the viewer, which is used to visualize the approximated surfaces in this chapter.

References

[Cha17] Charu, C. A. (2017). *Outlier analysis*. Springer. ISBN: 978-3-319-47578-3.

[Dok21] Dokken T., & Skytt, V. (2021). *SISL-SINTEF spline library, reference manual*, version 4.7. https://github.com/SINTEF-Geometry/SISL/

[Dok07] Dokken, T., & Skytt, V. (2007). Intersection algorithms and CAGD. In G. Hasle, K.-A. Lie, & E. Quak (Eds.), *Geometric modelling, numerical simulation, and optimization: Applied mathematics at SINTEF* (pp. 41–90). Springer.

[Haw80] Hawkins, D. (1980). *Identification of outliers*. Chapman and Hall.

[Mar18] Marussig, B., & Hughes, T. J. R. (2018). A review of trimming in isogeometric analysis: Challenges, data exchange and simulation aspects. *Archives of Computational Methods in Engineering*. https://doi.org/10.1007/s11831-017-9220-9

[Pan21] Pang, G., Shen, C., Cao, L., & Van Den Hengel, A. (2021). Deep learning for anomaly detection: A review. *ACM Computing Surveys*. https://doi.org/10.1145/3439950

[Pat02] Patrikalakis, N. M., & Maekawa, T. (2002). *Shape interrogation for computer aided design and manufacturing*. Springer.

[Rob19] Roberts, K. C., Lindsay, J. B., & Berg, A. A. (2019). An analysis of ground-point classifiers for terrestrial LiDAR. *Remote Sensing*. https://doi.org/10.3390/rs11161915

[Sky15] Skytt, V., Barrowclough, O., & Dokken, T. (2015). Locally refined spline surfaces for representation of terrain data. *Computers & Graphics*. https://doi.org/10.1016/j.cag.2015.03.006

[Sky16] Skytt, V., Dokken, T., Dahl, H. E. I., & Harpham, Q. (2016). Deconfliction, surface generation and LR B-splines. In M. Floater, T. Lyche, M.-L. Mazure, & K. Mørken (Eds.), *Mathematical methods for curves and surfaces*. 9th International Conference, MMCS 2016 Tønsberg, Norway, pp. 270–295. Springer Publishing Company.

[Sky22] Skytt, V., & Dokken, T. (2022). Scattered data approximation by LR B-spline surfaces. A study on refinement strategies for efficient approximation. In C. Manni & H. Speleers (Eds.), *Geometric challenges in isogeometric analysis* (Vol. 49). Springer INdAM Series.

[Sot06] Sotoodeh, S. (2006). Outlier detection in laser scanner point clouds. *The International Archives of the Photogrammetry, Remote Sensing and Spatial Information Sciences*. https://doi.org/10.3929/ETHZ-B-000037220

[Wan15]   Wang, Y., & Feng, H.-Y. (2015). Outlier detection for scanned point clouds using majority voting. *Computer-Aided Geometric Design, 62*, 31–43.

Chapter 6
LR B-Spline Surfaces and Volumes for Deformation Analysis of Terrain Data

Abstract Geospatial data acquisition of terrains with contact-free sensors such as Terrestrial or Airbone Laser Scanners generates scattered and noisy point clouds. Performing a surface approximation is an efficient way to reduce and structure the recorded point clouds. To that end, LR B-splines are attractive as they allow a local refinement, on the contrary to the tensor product B-spline and raster surfaces. By comparing the approximation error with a given tolerance, a local refinement is performed. We apply this adaptive refinement strategy to landslides data sets from Alpine terrain in Austria. We show how different epochs of the point clouds can be analyzed with LR B-spline volumes for spatio-temporal visualisation of deformation. We highlight the potential of a time-differenced LR B-splines volume for analysing geomorphological changes. A further application of this method is the drawing of contour lines.

Keywords GIS data set · Geospatial data set · LR B-splines · Adaptive surface fitting · Spatio-temporal deformation analysis · Contour lines · LR B-spline volumes · Geomorphological analysis

6.1 Introduction

Terrestrial Laser Scanners (TLS) are contact-free measuring sensors. They record dense point-clouds of objects or scenes by acquiring coordinates of points and an intensity value; this latter depends, e.g., on the reflected surface or atmospheric propagation, see Wujanz et al. [Wuj17]. TLS range measurements can be either based on phase shift or time-of-flight. We refer to Vosselman and Maas [Vos10] or Pfeifer and Briese [Phe07] for more details. Typically, the range measurement in phase shift is more accurate than time-of-flight TLS but the maximum range is smaller. Due to its high scanning rate, it is not uncommon that a point cloud contain millions of points, which need to be processed in some way. Well-known software are, e.g., the freely available CloudCompare [CC22], or MeshLab [Cig08]. Within a geodetic context, prominent applications using TLS point clouds are deformation analysis for objects such as tunnels [Jia21], dams [Gon08] or bridges [Zog08], see

© The Author(s) 2023
G. Kermarrec et al., *Optimal Surface Fitting of Point Clouds Using Local Refinement*,
SpringerBriefs in Earth System Sciences,
https://doi.org/10.1007/978-3-031-16954-0_6

also Mukupa et al. [Muk16]. Because TLS enables fast and precise mapping, they are currently used for monitoring forest canopy [Gri15] or landslides [Bar13].

The processing of huge point clouds can quickly become computationally unfavourable. Parametric spline surface approximation techniques address the challenge by reducing "data" into "information". The point clouds are efficiently compacted into a a manageable number of coefficients that are often estimated by least-squares (LS) adjustment. The mathematical modelization enables deformation analysis and rigorous statistical testing based on the estimated surfaces rather than on the original point clouds. Unfortunately, the Non Uniform Rational B-splines (NURBS, [Pie95]) do not allow for local refinement. This approximation method is unfavourable when the point clouds are scattered and noisy. Here the risk of overfitting should not be undertaken as it leads to unwanted ripples and oscillations in the approximated surface, see Bracco et al. [Bra18]. This latter may be confounded with deformations.

Starting from a first approximation of the point cloud using a coarse NURBS mesh, there exist three main approaches to perform an adaptive local refinement:

- Hierarchical B-splines (HB-splines) were introduced in Forsey and Bartels [For88]. The refinement is said to be dyadic and the cells of the mesh to be refined are halved at each iteration.
- T-splines are described, e.g., in Sederberg et al. [Sed03]. Here the refinement is performed by successively adding new control points in-between two adjacent control points in the T-mesh, Kermarrec and Morgenstern [Ker22] for an application to sand dunes point clouds from TLS.
- LR B-splines were developed by Dokken et al. [Dok13]. They are based on the concept of "splitting the B-splines", i.e., introducing new meshlines. Each meshline inserted has to split the support of at at least one tensor product (TP) B-spline. They were shown to be advantageous within GIS context, Skytt et al. [Sky15].

The remainder of this chapter is as follows: in a first step, we will shortly review the surface approximation with LR B-splines. The reader is referred to Chap. 2 and Chap. 3 for more details. We will introduce LR-B-spline volume as a promising tool for visualising and analysing spatio-temporal deformations. We will develop how contour lines can be computed from LR B-spline surfaces in a dedicated appendix. The chosen point cloud to illustrate the surface approximation is from the Alpine region in Austria.

6.2 Description of the Data Set

In the context of climate changes and the expansion of areas of urban settlement, e.g., in Alpine regions, early warning system for risk management necessitates high-quality data sets that are both spatially and temporally detailed. In this chapter, we perform spatio-temporal deformation analysis with mathematical approximations from a data set recorded in Austria, which will be shortly described in the following section.

Fig. 6.1 Left: View of the region under consideration, Right: Visualization of one point cloud with the software CloudCompare

6.2.1 Deformation Monitoring with a Terrestrial Laser Scanner

The data set studied in this chapter was recorded in the Valsertal region in Austria as part of HORIZON 2020 via the RFCS (Research Fund for Coal and Steel) funded research project i^2MON-"Integrated Impact MONitoring for the detection of ground and surface displacements caused by coal mining". Here we focus on the observations from the TLS (VZ-2000i, see http://www.riegl.com/nc/products/terrestrial-scanning/produktdetail/product/scanner/58/), Further information about the experiment can be found in dedicated publications, e.g., Schröder and Klonowski [Sch20].

The monitoring of the Valsertal aims to analyse deformation with the help of a long-range TLS, for underlying safeguard applications. A continuous series of measurements are available from August 13, 2020 up to and including September 8, 2020. In this contribution, we have selected a small excerpt from August 20, 2020 to August 22, 2020. We focus on an area in the lower part of an alpine slope where some rearrangements of the facilities took place in this period. The measurements were made in a refuge opposite an area affected by a rock fall as shown in Fig. 6.1. A geodetic instrument called a total station was installed on one of the two measuring pillars in the hut measuring 16 reflectors at a distance of approximately 250–800 m every hour. On the second pillar, a laser scanner was used for the permanent installation. One point cloud of the whole area was recorded every two hours resulting in 36 point clouds in total. In order to separate the expected apparent deformations within the time series from the influences of the georeferencing, the laser scanner was expanded to include a platform with inclination sensors in the horizontal plane of the local scanner coordinate system. The GNSS antenna on the scanner was used for time synchronization. This set up allows us to highlight the potential of surface approximation with LR B-splines from TLS observations with the aim to visualize spatio-temporal deformation based on mathematical surfaces and volumes.

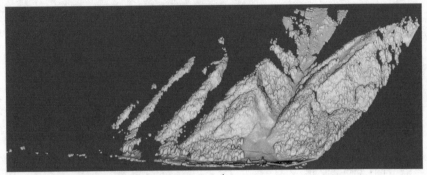

August 20th 1 AM

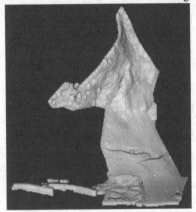

August 22th 1 AM   August 22th 11 PM

Fig. 6.2 Examples of the initial data sets

6.2.2 Data Set Preparation

The point clouds differ in size, depending on the epochs when they were recorded. Figure 6.2 illustrates the three different extensions of the selected point clouds. We have harmonized the point clouds so that they all fit with the smallest extension shown at the bottom right. The long tail to the left is excluded from the approximation. Obvious outliers are removed using the method of Chap. 5, Sect. 5.3.2.3. We note that the data sets from the first epochs include a large tree that later disappeared. In the later part of the period, the vegetation was filtered out to gain in accuracy. Figure 6.3 shows two examples of the point clouds processed for our study, with and without the tree. The tree was intentionally kept in the first point clouds to spice the spatio-temporal analysis. This way, we can show the potential of LR B-spline volume to detect such changes, with applications in forestry inventory, see, e.g., Liang et al. [Lia6].

<table>
<tr><td align="center">August 20th 11 AM</td><td align="center">August 21th 1 PM</td></tr>
</table>

Fig. 6.3 Examples of the processed data sets

6.3 Surface Approximation of the Selected Data Set

Approximating point clouds with tensor product (TP) B-spline surfaces does not allow for local refinement. LR B-splines are a way to locally refine the spline space and are shown to provide well-behaved mathematical surfaces for remote sensing applications, Skytt et al. [Sky22]. More specifically, an adaptive surface fitting is performed when the L1 norm—called in the following "error" or difference in absolute value—between the mathematical surface and the points inside a cell of the mesh exceeds a tolerance at a given step of the algorithm. The surface approximation obtained with LR B-splines depends on the tolerance from which the refinement is performed, the refinement method itself and the bidegree of the spline space. We have reviewed in Chap. 4 statistical methods to fix these parameters optimally.

6.3.1 General Principle of Adaptive Approximation Combining Least Squares and Multilevel B-Spline Approximation

We parameterize the data points by their x- and y-values, and approximate the z-values by an LR B-spline surface (function). The starting point of the iterative surface approximation with LR B-splines is a tensor product B-spline surface. Here a B-spline surface grid (also called also mesh) corresponds to the initial setting of the topology. Then an optimization is performed to compute the best LR B-spline surface corresponding to the initial mesh for approximating the point cloud. Once the initial surface is obtained, the error term between the mathematical surface and the points in the z-direction is computed. We consider the L1 norm, see Al-Subaihi et al. [AlS04]. If this value exceeds a given tolerance, a refinement is performed in the cells containing the corresponding points. The adaptive fitting is performed until a given number of iterations is reached or until no more error terms exceed the tolerance.

Extra degrees of freedom are inserted in the LR B-spline surfaces *locally*, where needed. The main advantage is that noise overfitting is avoided and the growth in data volume is limited, on the contrary to global fitting strategies where *all* cells are refined at *every* iteration steps. Clearly, with locally adaptive refinement methods, there will be cells that remain unchanged after a given iteration step because the error term is smaller than the chosen tolerance or there is no point in the cell. Intuitively, few cells are refined with a large tolerance whereas, all cells will be divided for a very low tolerance.

In the first iteration steps, the fitting is made using the LS method, i.e., the L2 norm or Euclidean distance between the parametrized point cloud and the parametric surface is minimised. As the number of refinement steps increases, it is favourable to switch to the Multilevel B-spline Approximation (MBA) developed by [Lee97], see also [Sky15] and Chap. 3 for more details on the procedure. For the MBA strategy, no equation system is solved as the coefficients are computed locally and explicitly: the residuals of the data points obtained from the last fitted surface are recursively approximated using finer meshes.

6.3.2   Approximation of the Selected Data Set

In this section, we present the results of the approximation of the domain under consideration, see Fig. 6.4.

6.3.2.1   Goodness of Fit

The goodness of fit is assessed using the usual criteria, as described in Chap. 3:

- The mean absolute distance abbreviated as MAE referring to the point cloud with respect to the approximated surface
- The maximum error Max_{err}
- The number of points outside tolerance n_{out}
- The number of coefficients n_{cp}
- The computational time CT. We used an 64-bit operating system with 8 GB RAM and an Intel(R) Core(TM) i5-63000U CPU @ 2.40 and 5.8 GHz.

6.3.2.2   Dealing with Outliers and Voids

Voids and outliers are typical challenges related to geospatial data, which will be shortly addressed in this section.

Voids

The point clouds contain voids which come from the scanning configuration and generate domains where no points are recorded. Data gaps challenge the approximation as the surface will try to extrapolate the known information in these areas, creating artificial oscillations. This drawback is particularly emphasized when LS approximation is used. MBA is more robust due to its explicit formulation and its similarity

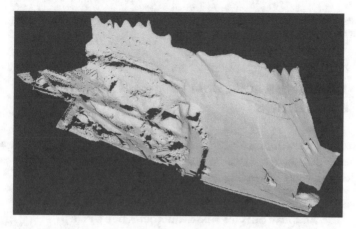

Fig. 6.4 Point cloud with about 1 million points acquired at 9 AM August 21st

with a L1 norm fitting, see Kermarrec and Morgenstern [Ker22] for a discussion. To face that challenge, we apply a 2-steps procedure:

- Restricting the values of the coefficients to an interval depending on the height range of the data points.
- Trimming: Trimming is a well known method within the context of Computed Aided Design, Medland and Mullineux [Med88]. Here the points are bounded by curves in the xy-plane. The curves are arranged in one loop for the outer boundary and one curve for each hole and associated to the parameter domain (xy-plane for points parameterized by their x- and y-values) of the surface. We refer to Chap. 5 for a detailed explanation.

Outliers

The point clouds under consideration contain noise and artifacts. These latter need to be removed prior to the surface approximation in order to avoid the fitting of outliers, which is unfavorable for further interpretation of the results.

Different methods exist to detect outliers. They are reviewed and applied in Chap. 5, Sect. 5.3.2.

Following strategies can be applied besides eliminating outliers:

- A smoothing term can be added to the function to minimize in the first step of the approximation [Sky15]. This way, smooth surfaces are obtained and ripples are avoided. Additionally, the MBA is known to be robust against outliers due to the use of the L1 norm of the error term to perform the adaptive refinement, Aigner and Jüttler [Aig07].
- It is possible to perform a classification previous to the fitting, as, e.g., in [Che118], Li et al. [Li16] or Xue et al. [Xue20] for a random forest classification algorithm. This way, objects such as trees or cars can be eliminated and only points corresponding to the ground approximated.

The data sets used in this chapter is cleaned for obvious outliers using the last procedure described in Chap. 5, Sect. 5.3.2.3.

6.3.3   Visualization of the Approximation at Different Iteration Steps

In this section, we will present some results of the surface approximation with the aim of being didactic and visual to improve the understanding of iterative surface approximation with LR B-splines.

6.3.3.1   The Adaptive Surface Approximation, Step by Step

A tolerance of 0.5 m is selected for the approximation of the selected point cloud. If a smaller tolerance is used, the risk of overfitting increases, which is unfavorable for a smooth and reliable surface fitting. A higher tolerance leads to a convergence after only few iterations, without the advantages of local refinement being visible, i.e., the approximation remains coarse.

We start the approximation with an initial biquadratic TP B-spline surface of 10 times 10 coefficients to approximate the point cloud. This corresponds to iteration 0 as shown in Fig. 6.5. Then the algorithm is allowed to run for 7 iterations, ending as shown in Fig. 6.12. The final surface output is shown in Fig. 6.12. We additionally draw the trimmed surface with respect to the point cloud domain (Fig. 6.13). Here the domains with no point are "cut" so that no unwanted effects such as, e.g., oscillations or drops occur in the domain without any observation.

6.3.3.2   Goodness of Fit

The results of the approximation are summarized in Table 6.1. Combined with Figs. 6.5, 6.6, 6.7, 6.8, 6.9, 6.10, 6.11 and 6.12, the impact of increasing the number of iteration steps is visible: after 5 iterations, the MAE does not decrease significantly (from 0.095 to 0.083 m), but the maximum distance does (from 0.095 to 0.083 m). This difference highlights the main advantage of local refinement: only specific domains are approximated locally with a higher accuracy by letting the main part of the point cloud untouched. The meshes allow to visualize more clearly the property of the iterative local refinement, although it is hardly possible to distinguish the differences in the last two iterations. Here the increase in the number of coefficients between iteration 5 and 7 (from 5700 to 14,000) is more descriptive (Table 6.1).

Figures 6.5, 6.6, 6.7, 6.8, 6.9, 6.10, 6.11 and 6.12 further highlight how the original coarse mesh from the first iteration is refined at each step. The number of meshlines, and so the number of coefficients, increases to account for local details *only*. Figure 6.13a shows where the points outside tolerance are located on the final surface after 7 iterations. The underlying final mesh is depicted additionally. There remain

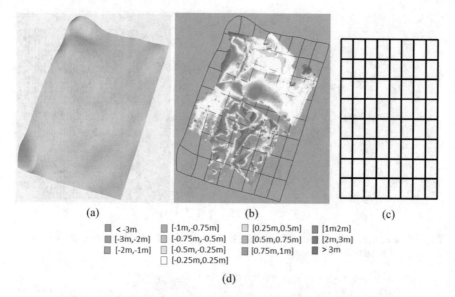

(a)   (b)   (c)

	< -3m		[-1m,-0.75m]		[0.25m,0.5m]		[1m2m]
	[-3m,-2m]		[-0.75m,-0.5m]		[0.5m,0.75m]		[2m,3m]
	[-2m,-1m]		[-0.5m,-0.25m]		[0.75m,1m]		> 3m
			[-0.25m,0.25m]				

(d)

Fig. 6.5 **a** Initial surface, **b** points coloured according to the distance to the surface, cell boundaries, **c** LR mesh (TP mesh), **d** colour scheme used for coloured point clouds in this figure and the subsequent figures

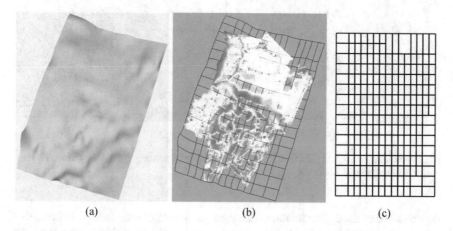

(a)   (b)   (c)

Fig. 6.6 **a** Surface after the first iteration, **b** points coloured according to the distance to the surface, cell boundaries, **c** LR mesh

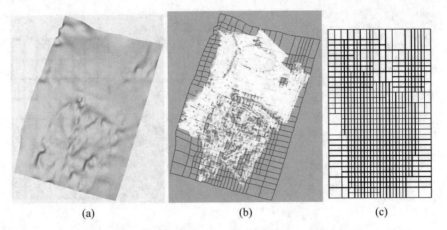

(a)   (b)   (c)

Fig. 6.7 **a** Surface after the second iteration, **b** points coloured according to the distance to the surface, cell boundaries, **c** LR mesh

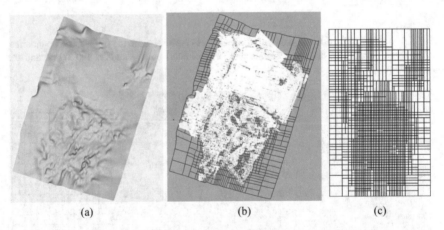

(a)   (b)   (c)

Fig. 6.8 **a** Surface after the third iteration, **b** points coloured according to the distance to the surface with cell boundaries, **c** LR mesh

domains where the approximation cannot be further improved: Table 6.1 shows that increasing the number of iterations does not lead to a smaller MAE after the sixth iterations. Indeed, the number of points outside tolerance decreases but in % of the total number of points, the difference is irrelevant (99.3 vs. 99.1% between the sixth and the sevenths iteration). At the same time, the number of coefficients increases from 9000 to 14,000. The computational time increases as well, but remains at a moderate level (a few sec). Thus, when searching for an optimal surface approximation, a balance has to be found between the number of iterations, the MAE and its relevance, the maximum distance and the computational time for a given tolerance. This choice is let to the practitioner which should judge the risk of fitting the noise as the number of iterations increases. An indication can be provided by searching the

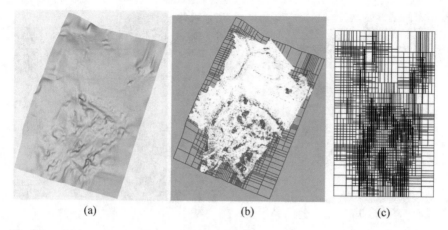

(a)   (b)   (c)

Fig. 6.9 a Surface after the fourth iteration, **b** points coloured according to the distance to the surface, cell boundaries, **c** LR mesh

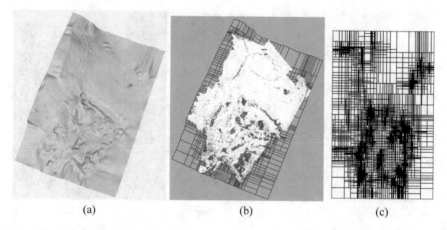

(a)   (b)   (c)

Fig. 6.10 a Surface after the fifth iteration, **b** points coloured according to the distance to the surface with cell boundaries, **c** LR mesh

minimum of AIC, as describe in Chap. 4. In our particular case, a minimum could be found after the eighth iteration. We point out that the AIC gives an indication about the turning point from which further refinement is not leading to a strong decrease of the root mean square error with respect to the increase of coefficients. It is a global criterion, that does not provide information about the local adjustment.

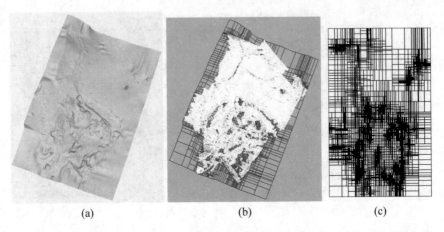

(a)   (b)   (c)

Fig. 6.11 **a** Surface after the sixth iteration, **b** points coloured according to the distance to the surface, cell boundaries, **c** LR mesh

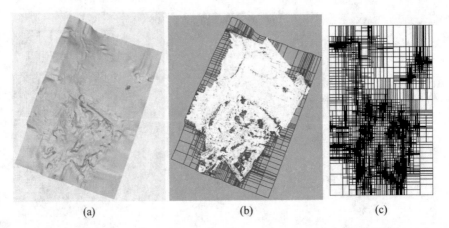

(a)   (b)   (c)

Fig. 6.12 **a** Final surface, **b** points coloured according to the distance to the surface, cell boundaries, **c** LR mesh

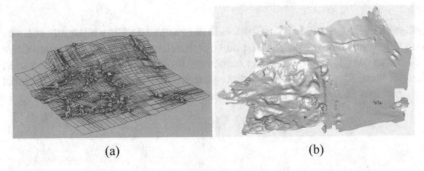

(a)   (b)

Fig. 6.13 **a** Points with a distance larger than the tolerance (0.5 m), **b** surface trimmed with respect to the point cloud domain (trimming with respect to the outer boundary only)

Table 6.1 Adaptive approximation of the point set shown in Fig. 6.4

Level	Max_{err}	MAE	n_{out}	n_{in} (%)	n_{cp}	CT
0	6.367	0.705	490,476	53.1	100	0.9
1	4.635	0.447	305,912	70.1	279	1.8
2	4.126	0.224	126,796	87.9	714	2.6
3	4.340	0.140	45,034	95.7	1866	3.5
4	3.942	0.111	24,111	97.7	3627	5.2
5	4.486	0.095	14,030	98.7	5737	6.0
6	4.482	0.086	9555	99.1	9242	7.0
7	3.917	0.083	7303	99.3	13,992	8.1

The maximum and average distance for each iteration level is reported along with the number of points with a distance larger than the tolerance, the percentage of resolved points and the number of surface coefficients. Spatial units are m and computation time is given in s

6.4 LR Spline Volumes to Analyse Spatio-temporal Deformation

The domain under consideration was scanned every two hours during three consecutive days. This leads to a large amount of point clouds, making the use of surface and *volume* approximations relevant to visualize and analyze the deformations or changes that may occur during that time. The main advantage is not having to work with the noisy and scattered point clouds. This is computationally advantageous and allows for a simpler interpretation.

6.4.1 Principle of Volume Approximation

To get an impression on the continuous development of the landscape in the selected area, the time component is added as a third parameter direction in the data set allowing an interval of 0.5 between each set in the time direction. Figure 6.14 shows the structure of the composed raw point clouds, before the volume approximation. The block of point clouds is narrow in the time direction compared to the space directions, and the time layers are distinguished by colour. The distances between these layers are larger than the distances between points in the xy-plane, but still small enough to control the behaviour of the spline volume approximation. The height range corresponding to the different epochs is very dependent of the presence of the aforementioned tree, i.e., the data set shown in Fig. 6.3a has a range of $[-42.69, 17.77]$ while the range for the points in (b) is $[-42.7, 8.55]$. The total height range in including all point sets is $[-42.71, 17.89]$.

A point cloud assembled from all epochs is approximated by an LR B-spline volume. Following Sect. 6.3.3, we apply a tolerance of 0.5 m and perform 7 iterations. Figure 6.16 shows the corresponding LR volume. The visualization is performed with

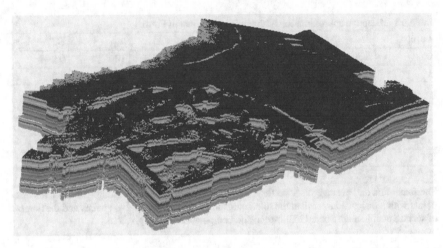

Fig. 6.14 The structure of the volume *point cloud*. The points are represented by their x-, y- and time coordinates and points from different acquisitions are distinguished by colours

a dedicated viewer as proposed in [Fuc17]. The colours are linked with elevation. Here we *do not* perform a surface approximation of each of the 36 point clouds independently and individually, as shown in the previous section: we approximate the block of point clouds as a whole.

The trivariate point cloud is approximated by an LR B-spline volume following the same approach as for surface approximation explained in Chap. 3.

1. The starting point is a TP B-spline volume, which is refined in an adaptive procedure.
2. The refinement is performed in a volumetric mesh cell when the distance between the value of the trivariate point and the LR B-spline volume exceeds a specified tolerance.
3. Then a mesh rectangle splitting at least one trivariate B-spline is inserted.

At each iteration step, an updated approximation using MBA is computed. The iteration stops when the given tolerance is met or a maximum number of iteration steps is applied. Figure 6.15 shows a mesh corresponding to a triquadratic LR B-spline volume with initially three inner knots in each parameter direction after one iteration. The refinement is performed only in the first parameter direction and the inserted mesh rectangles are highlighted with yellow colour.

We will refine in all three parameter directions simultaneously. The maximum distance between the point cloud and the volume after 7 iterations is 29 m. The most distant points are associated to the tree, which cannot be well fitted by a smooth surface. The average distance is 0.17 m and 1,507,346 out of 37,785,650 points have a distance to the volume larger than the tolerance of 0.5 m, meaning that 96% of the points are within the resolution. The maximum distance is kept at approximately the same level throughout the computation indicating a feature unsuitable to be fitted

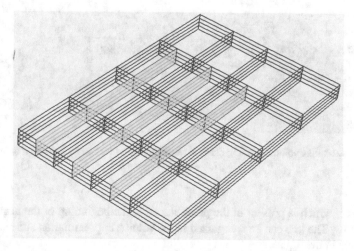

Fig. 6.15 Simple LR volume mesh with mesh rectangles

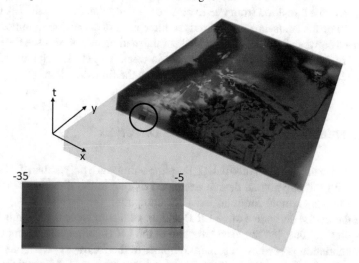

Fig. 6.16 LR B-spline volume approximating the trivariate point cloud. The height field is represented with colours

with a smooth volume, while the average distance is gradually reduced. The final number of coefficients is 93,829. The computational time is 10 min and 7 s excluding file operations, which is manageable from a practitioner perspective.

The domain of the LR B-spline volume approximation to the point cloud corresponds to the axis parallel bounding box surrounding the parameter points $\{x_i, y_i, t_i\}_{i=1}^{N}$ where N is the total number of points. As the boundaries of the point cloud do not adapt to axis parallel lines, for some parts of the volume shown in Fig. 6.16 there are no corresponding data points. The figure visualizes the height field corresponding to the point cloud with colours. The volume is cut at the position

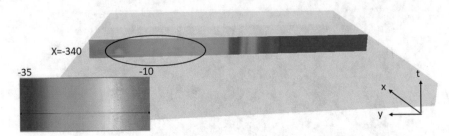

Fig. 6.17 LR-spline volume. Cutting plane in the x-direction gives some indication of landscape changes

of the tree, which is present in the point clouds at the beginning of the acquisition period only. The tree can be recognized in the volume approximation and is marked by a circle in the figure.

The presence or absence of the tree is the largest difference between the different data sets and the transition from tree to no-tree is totally dominant in the LR volume description of the landscape throughout the three days of acquisition. Some smaller differences between the data sets can be distinguished in the volume, see the marked area in Fig. 6.17. The corresponding marks is weak: to get clearer indications of change, we turn to the derivative of the volume in the time direction.

6.4.2   Volume Changes in the Time Direction

A partial derivative of a polynomial TP B-spline volume is a polynomial TP B-spline volume with the polynomial degree decreased by one in the direction of differentiation, and very simple formulas exist for computing the derivative. As a spline volume the partial derivative of an LR B-spline volume in some direction is also an LR B-spline volume, but the local structure of the LR B-spline volume implies that the differentiation procedure becomes complex. In the following, we represent, for each cell, the field as a spline volume without inner knots and differentiate cell by cell.

6.4.2.1   Partial Derivative of the Volume for Spatio-temporal Analysis

A non-zero derivative of the height field reveals changes in height: by focusing on the time direction only, we exclude landscape formations that do not change over time.

Figure 6.18 shows changes in time revealed by the time derivative of the LR B-spline volume. The visualization focuses on derivatives within a $[-3, 3]$ range. Higher values for the derivative field exist, but are not highlighted explicitly. The field is hidden for derivatives close to zero. We see that changes in the height field

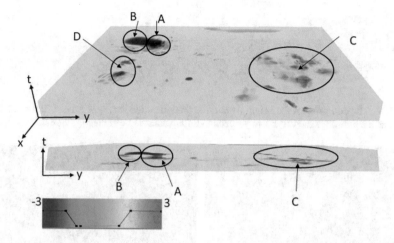

Fig. 6.18 The derivative of the height field in the time direction represented as an LR B-spline volume

Fig. 6.19 The derivative of the height field in the time direction just before August 20th at 11 PM (left) and just before 5 AM the same day (right)

occurs as blobs in the total volume. The figure presents an overview of the volume (top) and looks into it from the x-direction (bottom). The blobs of change are local in the time direction. They corresponds to (i) an object that appears and disappears or (ii) a modification of the landscape that is later left untouched. The spot marked with "A" represents the tree that is present in the first part of the time line. "B" is outside the point cloud and indicates that the height field is gradually decreased after removal of the tree. "C" and "D" are areas of interest to be discussed in the following.

Figure 6.19 focuses on the area marked with "C" in Fig. 6.18. Here it is marked with a circle. Green colour means no change in time, red means that the height increases while blue means that it decreases. The activities performed in period of time in this area are discussed in further detail in Sect. 6.4.3.

In Fig. 6.20 the area "D" is marked with an ellipse. In the cut to the right, we can see from the colours that the height field increases and then decreases again. In

Fig. 6.20 The derivative of the height field in the time direction just after August 21st at 11 PM (left) and a cut through the volume in y-direction hitting area "D" specified in Fig. 6.18

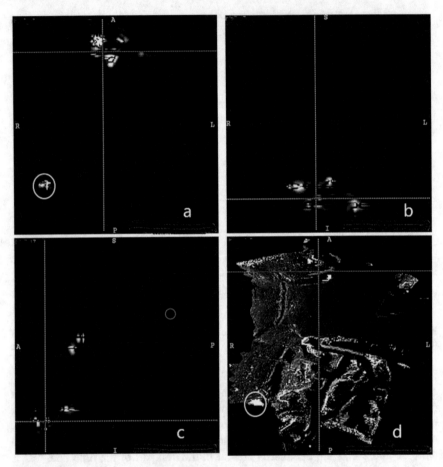

Fig. 6.21 The derivative of the height field in time direction at August 20th at 11 PM and corresponding differences in the point elevation, raster view

between, there is a narrow green strip indicating that the observed object, probably an excavator, is present. The incident to the right happens at an earlier point in time than the one to the left (right picture). This indicates a movement of the excavator from position "1" to position "2".

6.4.2.2 Spatio-temporal Changes Visualized as Medical Images

In Fig. 6.21, a visualization dedicated to medical images [ITK] is used to give an alternative view of the height changes. The volume is represented as a $255 \times 255 \times 255$ raster and the values are computed by averaging a number of sample points in the raster cells. The views in "a" and "d" show the same snapshot in time for different features of the volume. The derivative of the height field in the time direction for August 20th 11 PM is shown in Fig. 6.21, "a". Views "b" and "c" show cuts through the volume with constant y and constant x, respectively. The three views are connected by the blue cross. In "d" the average difference in point elevation for each cell is shown. The circle indicates the position of the tree in the views where it is present. Views "a" and "d" indicates that two objects of significant size are situated close to the blue cross and there are some smaller additional modifications of the elevation in the vicinity of the cross. The duration of the elevation changes can be seen from view "b" and "c" where the development in time is shown in the vertical direction. This reveals that the highlighted elevation changes were temporary. Note that a permanent change in elevation will appear only as a limited white spot in view "b" and "c". View "d" can be used to place an incidence in the landscape as the intensity map gives an indication of the terrain. White indicates steep areas or rapid landscape changes in time, while plains and areas without points are black.

Figure 6.22 provides another visualization of the situation in Fig. 6.20. The descent from the tree is completed as can be seen in view "b". Note that the derivative has its largest values when the volume adapts to a change and not when the peak or dump is at its largest. The time of the ring is when the tree is no longer present. The two positions of the excavator are shown in views "a" and "b". Due to the position of the cut, the appearance and disappearance of only one excavator are visible in "c".

6.4.3 Difference of Surfaces

In this section, we propose to analyse more specifically changes in the point cloud with LR B-splines surfaces. Here we look at the surface approximation of the point clouds acquired at 11 PM and 5 AM at August 20th, see Fig. 6.23. The maximum distance between the points and the surface is 28.58 m for (a) and 28.30 m for (b). The average distances are 0.178 and 0.179 m, and the number of points with a distance larger then 0.5 m is 25,583 and 24,801. The numbers of data points are 1,071,938 for (a) and 1,068,546 for (b). The main obstacle for an accurate approximation is the tree in the upper left corner, but also some excavators in the right half of the figures are impossible to represent exactly with a smooth surface. Some differences between Fig. 6.23a and b can be identified, but the general impression is that there is little difference in the landscape between the two epochs. Figure 6.24 reveals some more details. Here the difference surface between the two surfaces in Fig. 6.23 is computed and represented as an LR B-spline surface, *difference surface = surface b - surface a*. The surface is trimmed according to the point cloud at 11 PM. Contour

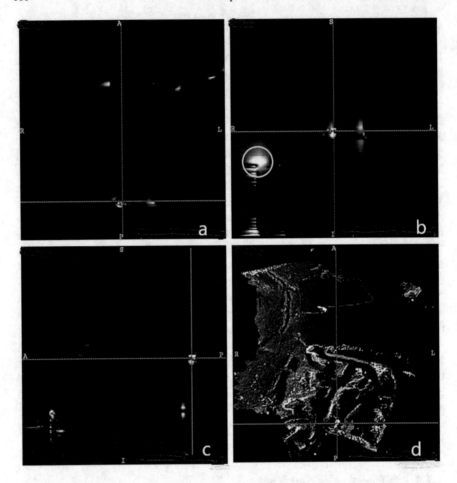

Fig. 6.22 The derivative of the height field in time direction at August 21st at 11 PM and corresponding differences in the point elevation, raster view

curves are computed for every 0.25 m between −5 and 5 m. Some details concerning the computation of these curves are given in Appendix 1. The green curves visualize material that is removed from (a) to (b) and red curves material that is added. The black curves show the zero level for the difference surface. Most of the surface is oscillating slightly below and slightly above zero. This is an effect of differences in the point clouds and the approximation error and does not represent a change in the landscape. The red and green curves in most cases represent change. At the point marked with "A" is the aforementioned tree and the difference here is not due to a real change. At "B", an excavator is added an at "C" one is removed. Snow is removed at "E" and moved to "D". Some local changes in the snow cover take place at "F". This short analysis highlights the potential of surface approximation to analyse deformation with application for geomorphological analysis. We refer to [And21] for an example based on the noisy and scattered point clouds.

(a)   (b)

Fig. 6.23 The situation at 11 PM (**a**) and 5 AM (**b**) at August 20th visualized as surfaces

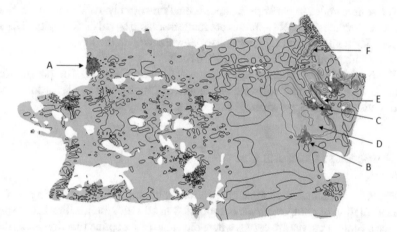

Fig. 6.24 The difference between the situation at 11 PM and 5 AM at August 20th visualized as a surface with associated contour curves for every 25th cm between −5 and 5 m, green curves represent negative levels, red curves positive and black the zero level

6.5 Conclusion

We have presented a local adaptive refinement strategy to approximate efficiently scattered and noisy point clouds from TLS. Prominent applications are deformation analysis or monitoring, without having to manipulate or filter a huge amount of data. To that end, we have used LR B-spline surfaces, which were shown to be well adapted to fitting terrains and seabeds. This approach refines the point clouds locally, avoiding the computation of unnecessary surface coefficients: the output is a compact surface in a short amount of time. This mathematical representation is favorable for further analysis of the point cloud; the noise is filtered out, and voids can be handled efficiently with CAD techniques such as trimming. The approximation method is based on a combination of LS, to which a smoothing term can be added in the first iteration steps, and MBA. Outliers are to be eliminated prior to the surface approximation. A classification can be performed in advance to eliminate, e.g., trees or cars if only the ground is of interest for deformation analysis.

We have applied the algorithm to TLS point clouds recorded in the Alpine region in Austria. The domain under consideration was scanned every two hours during three consecutive days. This large amount of data allows a visualization of change pattern from the mathematical approximations, without having to manipulate the original point clouds. To that end, we have introduced the LR B-spline volume and its derivative as a possibility to visualize spatio-temporal changes. The story of the point clouds could be guessed, paving the way for new applications of surface approximation within a GIS context. We have used images inspired by medical applications to visualize and analyse geomorphological changes. These examples highlight the potential of combining different visualization techniques to extract spatio-temporal information from a high number of point clouds.

The source codes to perform the approximation with bivariate (lrsplines2D) and trivariate (lrsplines3D) LR B-splines are made available by SINTEF Digital, Department of Mathematics and Cybernetics for downloading at the link: https://github.com/SINTEF-Geometry/GoTools.

The hardware requirements are Windows, Linux or MacOS. The program language is C++. Following software are required: Cmake, Boost, Qt for the viewer used to visualize the approximated surfaces in this chapter.

1   Appendix: Contour Curves

In Fig. 6.24, we showed contour curves corresponding to the underlying surface. The calculation of contour curves is supported in all GIS systems. For LR B-spline surfaces, contour curves are curves where the value of the spline function is constant.

To compute the contour curves, we search for curves $f_a(t) = (f_1(t), f_2(t))^T \in R^2$ such that $F(f_1(t), f_2(t)) = a$ for an LR B-spline surface F and an elevation value a. To that end, we split the LR B-spline surface into a number of TP B-spline surfaces. The division into TP B-spline surfaces is performed by a recursive algorithm. At each level, we consider how the current surface can be split by extending one meshline to cover the entire surface domain. The candidate meshline must contain T-joints, i.e., at least one meshline in the other parameter direction must end at this meshline. The number of surface elements overlapping the meshline extension should be minimized and at the same time the meshline should divide the current surface into two surfaces with roughly the same number of knots. The balance between the two criteria varies throughout the recursion levels. When an appropriate split is found, the algorithm proceeds to look for splits in the two sub-surfaces. The splitting algorithm stops when no sub-surface contains more meshlines that don't traverse the surface domain than a given threshold. Each sub-surface is expanded to a TP B-spline surface by adding missing mesh line segments.

Figure 6.25 illustrates the division of the difference surface presented in Sect. 6.4.3 into TP B-spline surfaces. Our aim is to study the computation of contour curves with zero height with some detail.

We use the interrogation functionality of SINTEF's spline library, SISL [Dok21] on each sub surface after the LR B-spline surface is split into TP B-spline surfaces.

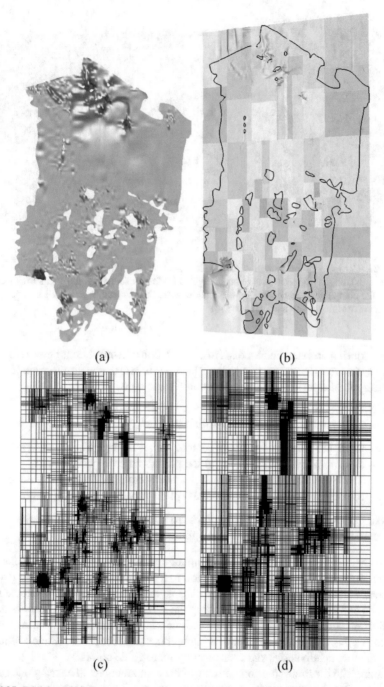

(a)

(b)

(c)

(d)

Fig. 6.25 Division of LR B-spline surface into a set of TP B-spline surfaces. **a** the initial surface, **b** the collection of TP B-spline surfaces distinguished by colour, the trimming curves corresponding to the LR B-spline surface are shown in black, **c** the LR mesh corresponding to the surface in **a**, **d** the mesh corresponding to the collection of TP B-spline surfaces

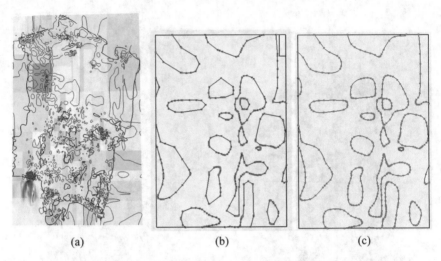

| (a) | (b) | (c) |

Fig. 6.26 Computation of contour curves. **a** Complete set of contour curves with elevation zero (blue curves), one TP B-spline surface is highlighted for further study, **b** guide points from the first part of the computation (red) with connections between them, **c** points generated by tracing the contour curves (green) and the final curves

The contouring problem corresponds to computation of intersections between a parametric spline surface and an algebraic surface, a problem that is discussed in [Pat02].

The applied algorithm can be divided into three parts:

1. Divide the LR B-spline surface into a set of TP B-spline surfaces
2. For each value a and each TP B-spline surface:

   (a) Compute the topology of the contour curves using SISL. This is a recursive algorithm that finds a set of "guide points" on each curve branch.
   (b) Trace each identified curve branch starting from an identified "guidepoint". Represent the curves traced out as spline curves.

3. For each value a, combine sub curves from different TP B-spline surfaces into contour curves for the entire LR B-spline surface.

An LR B-spline surface approximating an area with large shape variations will contain many details, which again will lead to a complex pattern of contour curves. Efficiency and robustness of the algorithm is reached through good interception methods and a clever strategy for dividing the surface into subsets. A discussion on subdivision strategies for surface intersections can be found in [Dok07]. A general rule is to subdivide at singularities and internal in closed loops. A complex situation leads to more subdivisions and consequently more guide points.

Figure 6.26 illustrates the computation of the contour curves. The red guide points in (b) are found at boundaries between sub surfaces. In such a complex situation, several recursion levels are required to be able to separate the different branches of the contours and ensure that no more closed contour curves exist. The last sub surface domain is shown in the upper right corner of the picture. All coefficients of the corresponding TP B-spline surface are negative. Thus, there is no possibility of a contour curve in this area and the computation can be finalized.

Given information about all contour curve branches in the area of interest, the curves can be drawn. Here the objective is to describe the curve with sufficiently accuracy, handle sharp turns in the curve and avoid jumping to a different contour curve. A marching procedure is applied. Given one point on the curve, a guess for the next point is made. The new point is moved to the contour curve and the segment between the two points is checked for consistency. The distance between the points is diminished if necessary. Figure 6.26c shows the tracing results. The density of the points is increased at sharp corners and when two curves pass within a small distance. Fragments of the contour curves are computed separately for each sub surface and the final step is to merge curve fragments across subset boundaries.

References

[Aig07] Aigner, M., & Jüttler, B. (2007). Robust fitting of parametric curves. *PAMM*. https://doi.org/10.1002/pamm.200700009

[AlS04] Al-Subaihi, I., & Watson, G. A. (2004). The use of the L1 and l∞ norms in fitting parametric curves and surfaces to data. *Applied Numerical Mathematics*. https://doi.org/10.1002/anac.200410004

[And21] Anders, K., Winiwarter, L., Mara, H., Lindenbergh, R., Vos, S. E., & Höfle, B. (2021). Fully automatic spatiotemporal segmentation of 3D LiDAR time series for the extraction of natural surface changes. *ISPRS Journal of Photogrammetry and Remote Sensing*. https://doi.org/10.1016/j.isprsjprs.2021.01.015

[Bar13] Barbarella, M., & Fiani, M. (2013). Monitoring of large landslides by Terrestrial Laser Scanning techniques: Field data collection and processing. *European Journal of Remote Sensing*. https://www.tandfonline.com/doi/abs/10.5721/EuJRS20134608

[Bra18] Bracco, C., Giannelli, C., Großmann, D., & Sestini, A. (2018). Adaptive fitting with THB-splines: Error analysis and industrial applications. *Computer Aided Geometric Design*. https://doi.org/10.1016/j.cagd.2018.03.026

[Che118] Chen, M., Pan, J., & Xu, J. (2018). Classification of terrestrial laser scanning data with density-adaptive geometric features. *IEEE Geoscience and Remote Sensing Letters*. https://doi.org/10.1109/lgrs.2018.2860589

[Cig08] Cignoni, P., Callieri, M., Corsini, M., Dellepiane, M., Ganovelli, F., & Ranzuglia, G. (2008). MeshLab: An open-source mesh processing tool. In *Sixth Eurographics Italian Chapter Conference* (pp. 129–136).

[CC22] CloudCompare (version 2.12) [GPL software]. (2022). Retrieved from http://www.cloudcompare.org/

[Dok07] Dokken, T., & Skytt, V. (2007). Intersection algorithms and CAGD. In *Geir. Hasle, Knut-Andreas. Lie et Ewald. Quak, coord.: Geometric modelling, numerical simulation, and optimization. Applied mathematics at SINTEF* (pp. 41–90), Springer, SINTEF.

[Dok13] Dokken, T., Pettersen, K. F., & Lyche, T. (2013). Polynomial splines over locally refined boxpartitions. *Computer Aided Geometric Design*. https://doi.org/10.1016/j.cagd.2012.12.005

[Dok21] Dokken, T., & Skytt, V. (2021). *SISL-SINTEF spline library, reference manual*, version 4.7. https://github.com/SINTEF-Geometry/SISL/

[For88] Forsey, D. R., & Bartels, R. H. (1988). Hierarchical B-spline refinement. In *SIGGRAPH 88 Conference Proceedings*, vol. 4, pp. 205–212.

[Fuc17] Fuchs, F. G., Barrowclough, O. J. D., Hjelmervik, J. M., & Dahl, H. E. I. (2017). *Direct interactive visualization of locally refined spline volumes for scalar and vector fields*. http://arxiv.org/abs/1707.01170

[Gon08] González-Aguilera, D., Gómez-Lahoz, J., & Sánchez, J. (2008). A new approach for structural monitoring of large dams with a three-dimensional laser scanner. *Sensors*. https://doi.org/10.3390/s8095866

[Gri15] Griebel, A., Bennett, L. T., Culvenor, D. S., Newnham, G. J., & Arndt, S. K. (2015). Reliability and limitations of a novel terrestrial laser scanner for daily monitoring of forest canopy dynamics. *Remote Sensing of Environment*. https://doi.org/10.1016/j.rse.2015.06.014

[ITK] ITK-SNAP. http://www.itksnap.org/pmwiki/pmwiki.php

[Jia21] Jia, D., Zhang, W., & Liu, Y. (2021). Systematic approach for tunnel deformation monitoring with terrestrial laser scanning. *Remote Sensing*. https://doi.org/10.3390/rs13173519

[Ker20] Kermarrec, G. (2020). *On estimating the hurst parameter from least-squares residuals*. Case study: Correlated terrestrial laser scanner range noise. *Mathematics*. https://doi.org/10.3390/math8050674

[Ker22] Kermarrec, G., & Morgenstern, P. (2022). Multilevel T-spline approximation for scattered observations with application to land remote sensing. *Computer-Aided Design*. https://doi.org/10.1016/j.cad.2022.103193

[Lee97] Lee, S., Wolberg, G., & Shin, S. Y. (1997). Scattered data interpolation with multilevel B-splines. *IEEE Transactions on Visualization and Computer Graphics*. https://doi.org/10.1109/2945.620490

[Li16] Li, Z., et al. (2016). A three-step approach for TLS point cloud classification. *IEEE Transactions on Geoscience and Remote Sensing*. https://doi.org/10.1109/tgrs.2016.2564501

[Lia6] Liang, X., Kankare, V., Hyyppä, J., Wang, Y., Kukko, A., Haggrön, H., et al. (2016). Terrestrial laser scanning in forest inventories. *ISPRS Journal of Photogrammetry and Remote Sensing*. https://doi.org/10.1016/j.isprsjprs.2016.01.006

[Med88] Medland, A. J., & Mullineux, G. (1988). *Principles of CAD. A coursebook*. Kogan Page. Online. https://ebookcentral.proquest.com/lib/kxp/detail.action?docID=3082883

[Muk16] Mukupa, W., Roberts, G. W., Hancock, C. M., & Al-Manasir, K. (2016). A review of the use of terrestrial laser scanning application for change detection and deformation monitoring of structures. *Survey Review*. https://doi.org/10.1080/00396265.2015.1133039

[Pat02] Patrikalakis, N. M., & Maekawa, T. (2002). *Shape interrogation for computer aided design and manufacturing*. Springer.

[Phe07] Pfeifer, N., & Briese, C. (2007) Laser scanning—Principles and applications. In *GeoSiberia 2007—International Exhibition and Scientific Congress. European Association of Geoscientists and Engineers*. https://doi.org/10.3997/2214-4609.201403279

[Pie95] Piegl, L. (1995). *The NURBS book*. Springer. ISBN: 978-3-642-97385-7.

[Sch20] Schröder, D., & Klonowski, J. (2020). i MON–Integriertes monitoring von Oberflächen- und Untergrundbewegungen verursacht durch den Kohlebergbau. *Ingenieurvermessung* (pp. 19–20). Internationaler Ingenieurvermessungskurs.

[Sed03] Sederberg, T. W., Zheng, J., Bakenov, A., & Nasri, A. (2003). T-splines and T-NURCCs. *ACM Transactions on Graphics*. https://doi.org/10.1145/882262.882295

[Sky15] Skytt, V., Barrowclough, O., & Dokken, T. (2015). Locally refined spline surfaces for representation of terrain data. *Computers & Graphics*. https://doi.org/10.1016/j.cag.2015.03.006

[Sky22] Skytt, V., & Dokken, T. (2022). Scattered data approximation by LR B-spline surfaces. A study on refinement strategies for efficient approximation. In C. Manni & H. Speleers (Eds.), *Geometric challenges in isogeometric analysis* (Vol. 49). Springer INdAM Series.

[Vos10] Vosselman, G., & Maas, H.-G. (2010). Airborne and terrestrial laser scanning. Whittles Publishing (Distributed in North America by CRC).

[Wuj17] Wujanz, D., Burger, M., Mettenleiter, M., & Neitzel, F. (2017). An intensity-based stochastic model for terrestrial laser scanners. *ISPRS Journal of Photogrammetry and Remote Sensing*. https://doi.org/10.1016/j.isprsjprs.2016.12.006

[Xue20] Xue, D., Cheng, Y., Shi, X., Fei, Y., & Wen, P. (2020). An improved random forest model applied to point cloud classification. In *IOP Conference Series: Materials Science and Engineering*. https://doi.org/10.1088/1757-899X/768/7/072037

[Zog08] Zogg, H. M., & Ingensand, H. (2008). Terrestrial laser scanning for deformation monitoring: Load tests on the Felsenau Viaduct (CH). *The International Archives of the Photogrammetry, Remote Sensing.* https://doi.org/10.3929/ETHZ-B-000011210

Chapter 7
Conclusion

In this SpringerBrief, we went through the mathematical concepts of LR B-splines in all its facets, explaining in details how the spline space can be refined and the different strategies for adaptive approximation. Through detailed example with seabed and terrain data sets, we have highlighted how adaptive surface approximation of various noisy and scattered point clouds is performed concretely. We showed how to deal with challenges raised when working with real data set such as voids or outliers. We presented numerous applications that can be derived from "transforming data to information". More specifically, we reviewed:

1. How *adaptive local refinement* can be performed by combining multi-level B-spline approximation and least squares with a low computational burden,
2. How parameters such as tolerance, number of iterations of the algorithm, and refinement strategies affect the surface approximation,
3. How *statistical concepts* can be used to judge the goodness of fit and determine parameters of the surface approximation, such as the tolerance, or the bidegree of the splines space,
4. How *outliers* can be removed efficiently and how to *fuse* data from different sources to perform efficient surface approximation,
5. How *voids* can be handled by applying trimming,
6. The potential of LR volumes for spatio-temporal analysis of point clouds,
7. The computation of contour lines from the mathematical approximations as an additional application.

We have illustrated the principle of surface approximation with various examples, using data from terrestrial laser scanner, sonar, and terrain or seabed data set. We have proposed the LR B-spline surface and volume as a new and promising format for representing noisy and scattered point clouds in a compact form. This approximation method provides a middle road between the rigid, but effective regularity of the raster format and the large flexibility of triangulated surfaces. LR B-spline surfaces are smooth and can, due to their adaptive potential, represent local detail without a drastic increase in data size. For point clouds coming from sensors having a very

© The Author(s) 2023
G. Kermarrec et al., *Optimal Surface Fitting of Point Clouds Using Local Refinement*,
SpringerBriefs in Earth System Sciences,
https://doi.org/10.1007/978-3-031-16954-0_7

high data rate and containing millions of points, the computational time of the surface approximation stays manageable: This is a strong argument for a wide acceptance of the local iterative fitting to represent scattered and noisy data mathematically. The surfaces can be exported as rasters in various resolutions as well as collections of tensor product spline surfaces. All software are freely available to promote usage.

7.1   A Promising Application: The LR B-Spline Volume for Spatio-temporal Analysis of Geomorphological Changes

The analysis of spatio-temporal deformation is one of the most promising applications of LR B-spline surfaces and volumes. These latter will provide a framework for analysing geomorphological changes without having to work on noisy and scattered point clouds of different quality, coming from different sensors. Detecting and analysing the movements of, e.g., sand dunes and snow masses is a potential use of LR B-spline volumes to gain a better understanding in the underlying geophysical processes. Similar applications are conceivable within the context of geodetic deformation analysis of structures, such as tunnels or dams. Advanced methods could take features such as extremal points, ridges and valleys into account. Outliers should be efficiently identified during the computation of the approximating surface and removed during the iterative algorithm presented. The outlier problematic is an important topic for further work and has been addressed in the present SpringerBrief.

7.2   Ongoing Research

Ongoing research focuses on identifying the most optimal surface. New criteria could be investigated to judge the quality of the approximation by accounting for the sensor noise, setting a tolerance and a stop criterion for the number of iterations in the approximation algorithm. We have introduced this challenge within the context of model selection, but alternative strategies should be developed. They could depend on the application at hand, as a balance between computational time, needed accuracy of the surface and overfitting avoidance. The detection and analysis of deformation or changes between two or more surface approximations or inside an LR B-spline volume remain an open topic, for which the definition of distance between mathematical surfaces should be further investigated. This latter should account for both the uncertainty of the measurements and the surface/volume approximation.

Last but not least, we expect that the capacity of LIDAR and sonar type data acquisition technology will continue to grow in the next decades. This makes in even more urgent to turn the enormous more or less structured point clouds into mathematically structured information that can be efficient interrogated and analysed.

We believe that representations that allow the granularity of the representation to adapt to the local behaviour of the underlying information intrinsic in point clouds are essential in this respect.

Printed in the United States
by Baker & Taylor Publisher Services